RAIDING THE SANCTUARY

Redcatchers in Cambodia,
May 12th - June 25th, 1970

Robert J. Gouge

Bloomington, IN Milton Keynes, UK

authorHOUSE

AuthorHouse™
1663 Liberty Drive, Suite 200
Bloomington, IN 47403
www.authorhouse.com
Phone: 1-800-839-8640

AuthorHouse™ UK Ltd.
500 Avebury Boulevard
Central Milton Keynes, MK9 2BE
www.authorhouse.co.uk
Phone: 08001974150

First published by AuthorHouse 5/9/2006

ISBN: 1-4259-3134-0 (sc)

Printed in the United States of America
Bloomington, Indiana

This book is printed on acid-free paper.

CONTENTS

DEDICATION

"To The Good Guys"

Sp4 Richard G. Desillier B/5-12, 5-13-70
SGT Arlie "Pete" Spencer Jr. B/5-12, 5-15-70
Sp4 Robert J. Urbassik B/5-12, 5-19-70
Sp4 Donald G. Busse C/5-12, 5-21-70
WO1 Robert E. Gorske HHC-199th, 5-21-70
SGT John W. Rich C/5-12, 5-21-70
Sp5 Warren L. Scanlan Jr. C/5-12, 5-21-70
PFC Daniel E. Nelms D/2-40, 5-22-70
Sp4 Dannie L. Hawkins B/5-12, 5-29-70
Sp4 Frederick R. Levins 76th CTT, 6-16-70
SGT Michael W. Notermann D/5-12, 6-19-70
PFC Ronald R. Stewart D/5-12, 6-19-70
PFC Johnny M. Watson D/5-12, 6-19-70
PFC Charles C. Cisneros D/5-12, 6-22-70
PFC Raul De Jesus-Rosa D/5-120, 6-22-70
PFC Allen E. Oatney D/5-12, 6-22-70

Your lives, memories, deeds and sacrifices will always be remembered fondly by your friends and family. You were once living, breathing, happy young men in the prime of youth, yet you willingly went, served and gave your lives in Vietnam and Cambodia when so many others in your generation did not. The pain of your loss will always be felt, but you will never be forgotten. This book is dedicated to you and to your families and loved ones.

THE CHARGE OF THE 199TH LIGHT INFANTRY BRIGADE

by Lord Alfred Tennyson
(with a few adjustments by Bob Fromme, D/4-12)

Half a click, half a click,
Half a click onward,
All in the Delta, the valley and jungle of Death
Humped the gallant Redcatchers.
They humped the boonies with their "iron pigs,"
their M16s, M79s, mortars and more.
'Forward, the 199th Light Infantry Brigade!

Charge for the guns!' he said:
Into the Delta paddies and jungles of Death
Though the bamboo, the knee-deep paddy mud,
Humped the Redcatchers.
'Forward, the 199th Light Infantry Brigade!'
Was there a man dismay'd ?
Not tho' the soldier knew
The Nation's leaders had no courage, no resolve.
More then one politician had blunder'd:

But for the Redcatchers,

Their's not to make reply,
Their's not to reason why,
Their's but to do and die:
Into the Delta, the valley and jungles of Death
Humped the Gallant Redcatchers.

Rockets to right of them,
Mortars to left of them,
ChiComs in front of them
AK 47s volley'd and thunder'd;
Storm'd at with shot and shell,
Booby Trapped and Rocketed as well,
Tired, Frightened and Bloody,
But boldly they humped on and well,
Into the jaws of Death,
Into the mouth of that Southeast Asian Hell
Humped the Gallant Redcatchers.

Flash'd all their bayonets bare,
Flash'd as they turn'd in air
Wasting a few "Charlies" there,
Charging an army, while
Hanoi Jane played the media for fame, while
Flower children danced in the Woodstock rain, while
Bankers and Politicians counted their change,
All the world wonder'd:

Plunged in the battery-smoke
Right thro' the line they broke:
Valiant young Americans:
Kentuckian, Californian, Puerto Rican and more
Reel'd from the machete-stroke
Shatter'd and sunder'd.

They bandaged their wounded,
Carried too many stretchers,
They filled too many body bags.

Too lonely, too exhausted, too frightened to grieve,
Humped on through the paddies, the jungles of Death
Humped on into the days, weeks, years,
the Gallant Redcatchers.
Then the remaining rode
The "Freedom Bird" back, but not
Not the number that were sent.

Flower children to right of them,
Hecklers to left of them,
Protesters behind them
Volley'd and thunder'd;
Storm'd at with shout,
With whistle and Yell,
All this after so many young Americans,
Young warriors fell,
They that had fought so well
Those that remained,
Those who came thro' the jaws of Death,
Back from the mouth of Hell,
All that was left of them,
Left of the Redcatchers.

When can their glory be known?
O the wild charges they made!
All the world wonder'd, wanted to forget
And so they went on.
Quietly the Redcatchers faded,
Faded into whatever future they could make.
Whatever life they could find beyond the pain,
The cold dark memories of the Delta, the valley and jungles of Death
When will the world Honor the charges they made?
Who will be there to honor the 199th Light Brigade?

Those noble Redcatchers!

ACKNOWLEDGEMENTS

These men, these former Redcatchers are truly amazing. This is my third work on the 199th Light Infantry Brigade and each time in each endeavor, veterans of the Brigade have, without hesitation, offered to help me time and time again in any way possible. For this book, several of men have graciously loaned me pictures, letters, documents and personal information, even though I have never met them in person. That speaks volumes about their character, integrity and honesty. To an outsider such as myself, I think that gesture is awesome. We have all agreed that even though this story is painful, it is one that needs and deserves to be told, not only for them, but for their families, friends and for history as well. I would like to give a respectful and a heartfelt "Thank You" to the following men for their time, memories, stories and friendship. It is men like these that have made America great and carry on the 199th Light Infantry Brigade legacy. Just like the brave and noble 58,000 men and women that gave their lives for their country in the Vietnam War, they too are heroes and role models in their own right, whether they realize it or not.

David Ashworth, HHC 5-12
Terry Braun, B/5-12
Walter Case, B/5-12
David Cook, A/5-12
Ihor Dopiwka, E/5-12
John Hart, D/5-12
Rolf Hernandez, A/5-12

Stan Hogue, D/2-40
Jim Horine, B/5-12
Bob Kenna, B/5-12
Roger Lowery, HHC 5-12
Albert Malone, CO 5-12th Infantry
Reid Mendenhall, C & E 5-12
Kay Moon, D/5-12
Ron Orem, B/5-12
Bert Ovitt, HHC 5-12
Bob Pempsell, D/5-12 & D/2-40
William Rose, A/5-12
Kevin Scanlon, A/5-12
Robert Schwaber, D/2-40
Malcolm Smith, B/5-12
Allen Thomas, B/5-12
John Wensdofer, C/5-12

Special thanks to Larry McDougal (A/2-3, 1967-1968) the chief historian of the 199th LIB's Redcatcher Association and his wife Pat for their time, friendship, advice and generous copies of the Cambodian after-action reports in their entirety.

FORWARD

This book deals strictly with the actions of the 5th Battalion, 12th Infantry, Delta Battery, 2nd Battalion, 40th Artillery, Fireball Aviation, Company M, 75th Infantry (Ranger) and the 76th Combat Tracker Team Detachment of the 199th Light Infantry Brigade in the Cambodian Incursion from May 12th to June 25th, 1970. It may be said by the reader that there is a bias here. There is. These men from the 199th Light Infantry Brigade were to initially be held in reserve for the 1st Cavalry Division (Airmobile) and used at their discretion when needed. Little did the rank and file of the Warrior battalion know that during their two month deployment in Cambodia, they would see heavy and near constant combat with some of the most determined and toughest soldiers in the North Vietnamese Army.

Yet history and historians of the Vietnam War have often overlooked or neglected to make aware the pivotal role that these light infantry units played during that highly controversial operation in the summer of 1970. Some very recent works on the subject even state that, "The 199th LIB played no direct role in the Incursion." Other facts about the Brigade in Cambodia are either incorrect, misrepresented or downplayed by the larger units that were also there in the campaign during that time. Nothing could be more damaging to the Brigade's excellent history or further from the truth!

During that harrowing and strenuous mission, approximately 590 men from the 5-12th Infantry and its attached units were deployed to Cambodia. By June 25th, the battalion had lost 16 soldiers killed in action. Another seventy-nine were wounded, along with countless others who will carry the terrifying memories of Cambodia with them for the rest of their lives.

Not only did these Redcatchers from the 199th Light Infantry Brigade participate in some of the heaviest firefights of the Cambodian Incursion, they also found, captured and destroyed some of the largest enemy supply caches of the Vietnam War. Their contribution and experience cannot be underestimated. This is their story, told by the veterans that fought there. May it never be forgotten and long live the memory of the men from the 199th LIB.

INTRODUCTION

By the spring of 1970, the Communist North Vietnamese and their southern Viet Cong allies had used the relative safety of the Cambodian border regions with South Vietnam for well over five years. Their use of this largely remote and supposedly "neutral" area to store supplies, train, rest and re-equip their troops coming down the Ho Chi Minh Trail had seriously hampered the United States' efforts to stop the flow of Communism into South Vietnam. In short, the use of Cambodian sanctuaries by the North Vietnamese and Viet Cong proved to be a mortal blow to the United States and the Republic of South Vietnam.

Once a colony of France, Cambodia had been in turmoil, along with most other countries in Southeast Asia, since the French defeat by the Viet Minh at Dien Bien Phu in 1953. While North Vietnam was taken over by the Communists, the decades old royal family returned to power in Cambodia. When the North Vietnamese began marching southward in the mid-1950's and early 1960's, Prince Norodom Sikanouk sought to keep his country neutral by trying to appease the United States, China, the Soviet Union and both North and South Vietnam. At this expense, the forces of North Vietnam were able to use the border areas with South Vietnam as long as they did not incite or give military support to the pro-communist Khmer Rouge.

Because of Prince Sikanouk's politics, "The North Vietnamese saw an opportunity to use the Cambodian border area with South Vietnam, Cambodian neutrality, and bribery to turn Cambodia into a logistics base for Communist forces in South Vietnam."

Prince Sihanouk's reign did not last for long however. In March of 1970, he was ousted from power by Lon Nol, a general in the Cambodian Army and the acting prime minister who vowed to chase the North Vietnamese out of the country.

By late 1969 and early 1970, "Although the North Vietnamese were not openly using the ports and roads from the sea to their jungle hideouts, they were hiring Cambodian firms to do it. Ships from Communist countries were regularly unloading food and ammunition at the Cambodian ports. Cambodian trucking companies then drove off with their stuff, and it ended up in North Vietnamese depots in the jungle, three miles from the South Vietnamese border." By doing this, the North Vietnamese were supplying more than 200,000 Communist troops in South Vietnam alone and nearly all of their operations in the Mekong Delta.

Throughout 1965, 1966 and 1967, the United States secretly dropped thousands of tons of bombs on the neutral country. By 1968, especially after the countrywide Tet Offensive where the defeated but surviving NVA and VC forces limped back to their Cambodian sanctuaries to regroup and reorganize, it was clear to both the military hierarchy in Vietnam and President Richard M. Nixon that the border area would have to be attacked by Allied ground units. By the end of that year, the groundwork for the Cambodian Incursion was being laid.

In the latter months of 1969, various Army units, namely the 1st Cavalry Division (Airmobile) and various ARVN units began redeploying and operating in areas closer to the Cambodian/South Vietnamese border.

In mid-March, 1970, while enemy contacts were at low to moderate levels in the III Corps Tactical Zone, MACV (Military Assistance Command, Vietnam) began meeting and discussing about the feasibility of a limited ground strike into Cambodia and on April 24th, 1970, the commanding general of II Field Force was given the green light to go ahead and begin preparing for an offensive attack against the vast Communist depots and base areas in the infamous "Fishhook" region.

Codenamed "Task Force Shoemaker," the assault force consisted of the 1st Cavalry Division, 11th Armored Cavalry Regiment, 4th Infantry Division, 25th Infantry Division, 3rd Brigade, 9th Infantry Division and the 5th Battalion, 12th Infantry and Delta Battery, 2nd Battalion, 40th Artillery, 199th Infantry Brigade (Sep)(Lt).

The major goal of the strike into Cambodia was to be the total destruction of the Communists' cross-border facilities, supplies and material. At the time of the planned attack, there was an estimated 27,000 enemy troops in the region. However, based on captured documents and prisoners taken during the first days of the operation, this changed to a rather conservative estimate of well over 63,000 enemy troops in the Fishhook alone, when the American pilots, grunts, engineers and tankers crossed the border.

As well as Operation Toan Thang (as the Incursion was officially called) was received by both MACV and President Nixon, there were severe flaws in the plan. The most obvious and damaging was the decision to limit the "blitzkrieg" no more than 30 kilometers into the country, along with the promise that all U.S. forces would withdraw completely from the country in exactly two months. In short, all the Communists had to do was pull back beyond the boundary line and wait for the deadline. Another negative aspect of Operation Toan Thang was the American public's reaction to news of the decision to raid the sanctuaries, especially when President Nixon made the announcement to do so on national television just hours after the action had already started.

When Task Force Shoemaker thundered across the border in the early morning hours of May 1st, 1970, the U.S. soldiers had very nearly achieved a complete tactical surprise on the North Vietnamese and Viet Cong. It was not until May 12th, when the NVA attacked elements of the newly-arrived 5-12th Infantry at LZ Brown, that they began mounting any serious resistance or counter-attacks. Despite this surprise, however, the Communists were still able to move an estimated 400-600 tons of supplies out of the danger zone both before and during the first days of the Incursion.

For the next forty-five days, there was not only near sustained and heavy fighting between the North Vietnamese and units of the 5-12th Infantry, but with all American ground units participating in the campaign as well. In the end, the Cambodian Incursion of 1970 was a huge success, although most of the major American media and news agencies, just like the Tet Offensive of 1968, did not report it as such.

The Incursion severely damaged the Communist logistical and support bases in Cambodia and Vietnam, thus disrupting Hanoi's plans for the next two years. Over 11,000 North Vietnamese were killed or wounded. It also decreased the number of casualties that the United States would have suffered while pulling out of Vietnam during the Vietnamization Program. The ammunition found and destroyed alone, could have supplied most, if not all of the Viet Cong and North Vietnamese soldiers in South Vietnam for well over 10 months and the number of weapons captured would have been issued to over 74 NVA Infantry battalions. The rice destroyed could have fed more than 25,000 enemy soldiers for the year.

It can also be argued that because of the cross-border offensive, the downfall of Saigon was postponed years later until 1975. The vast majority of the American troops that served in Cambodia knew, both then and now, that Nixon's decision to go ahead with the Incursion was right. Time has shown this to be true.

Good evening my fellow Americans

Ten days ago, in my report to the Nation on Vietnam, I announced a decision to withdraw an additional 150,000 Americans from Vietnam over the next year. I said then that I was making that decision despite our concern over increased enemy activity in Laos, in Cambodia, and in South Vietnam.

At that time, I warned that if I concluded that increased enemy activity in any of these areas endangered the lives of Americans remaining in Vietnam, I would not hesitate to take strong and effective measures to deal with that situation.

Despite that warning, North Vietnam has increased its military aggression in all these areas, and particularly in Cambodia.

After full consultation with the National Security Council, Ambassador Bunker, General Abrams, and my other advisers, I have concluded that the actions of the enemy in the last 10 days clearly endanger the lives of Americans who are in Vietnam now and would constitute an unacceptable risk to those who will be there after withdrawal of another 150,000.

To protect our men who are in Vietnam and to guarantee the continued success of our withdrawal and Vietnamization programs, I have concluded that the time has come for action.

Tonight, I shall describe the actions of the enemy, the actions I have ordered to deal with that situation, and the reasons for my decision.

Cambodia, a small country of 7 million people, has been a neutral nation since the Geneva agreement of 1954 - an agreement, incidentally, which was signed by the Government of North Vietnam.

American policy since then has been to scrupulously respect the neutrality of the Cambodian people. We have maintained a skeleton diplomatic mission of fewer than 15 in Cambodia's capital, and that only since last August. For the previous 4 years, from 1965 to 1969, we did not have any diplomatic mission whatever in Cambodia. And for the past 5 years, we have provided no military assistance whatever and no economic assistance to Cambodia.

North Vietnam, however, has not respected that neutrality.

For the past 5 years - as indicated on this map that you see here - North Vietnam has occupied military sanctuaries all along the Cambodian frontier with South Vietnam. Some of these extend up to 20 miles into Cambodia. The sanctuaries are in red and, as you note, they are on both sides of the border. They are used for hit and run attacks on American and South Vietnamese forces in South Vietnam.

These Communist occupied territories contain major base camps, training sites, logistics facilities, weapons and ammunition factories, airstrips, and prisoner-of-war compounds.

For five years, neither the United States nor South Vietnam has moved against these enemy sanctuaries because we did not wish to violate the territory of a neutral nation. Even after the Vietnamese Communists began to expand these sanctuaries 4 weeks ago, we counseled patience to our South Vietnamese allies and imposed restraints on our own commanders.

In contrast to our policy, the enemy in the past 2 weeks has stepped up his guerrilla actions and he is concentrating his main forces in these sanctuaries that you see on this map where they are building up to launch massive attacks on our forces and those of South Vietnam.

North Vietnam in the last 2 weeks has stripped away all pretense of respecting the sovereignty or the neutrality of Cambodia. Thousands of their soldiers are invading the country from the sanctuaries; they are encircling the capital of Phnom Penh. Coming from these sanctuaries, as you see here, they have moved into Cambodia and are encircling the capital.

Cambodia, as a result of this, has sent out a call to the United States, to a number of other nations, for assistance. Because if this enemy effort succeeds, Cambodia would become a vast enemy staging area and a springboard for attacks on South Vietnam along 600 miles of frontier - a refuge where enemy troops could return from combat without fear of retaliation.

North Vietnamese men and supplies could then be poured into that country, jeopardizing not only the lives of our own men but the people of South Vietnam as well.

Now confronted with this situation, we have three options.

First, we can do nothing. Well, the ultimate result of that course of action is clear. Unless we indulge in wishful thinking, the lives of Americans remaining in Vietnam after our next withdrawal of 150,000 would be gravely threatened. Let us go to the map again. Here is South Vietnam.

Here is North Vietnam. North Vietnam already occupies this part of Laos. If North Vietnam also occupied this whole band in Cambodia, or the entire country, it would mean that South Vietnam was completely outflanked and the forces of Americans in this area, as well as the South Vietnamese, would be in an untenable military position.

Our second choice is to provide massive military assistance to Cambodia itself. Now unfortunately, while we deeply sympathize with the plight of 7 million Cambodians whose country is being invaded, massive amounts of military assistance could not be rapidly and effectively utilized by the small Cambodian Army against the immediate threat. With other nations, we shall do our best to provide the small arms and other equipment which the Cambodian Army of 40,000 needs and can use for its defense.

But the aid we will provide will be limited to the purpose of enabling Cambodia to defend its neutrality and not for the purpose of making it an active belligerent on one side or the other.

Our third choice is to go to the heart of the trouble. That means cleaning out major North Vietnamese and Vietcong occupied territories these sanctuaries which serve as bases for attacks on both Cambodia and American and South Vietnamese forces in South Vietnam. Some of these, incidentally, are as close to Saigon as Baltimore is to Washington. This one, for example, is called the Parrot's Beak. It is only 33 miles from Saigon.

Now faced with these three options, this is the decision I have made.

In cooperation with the armed forces of South Vietnam, attacks are being launched this week to clean out major enemy sanctuaries on the Cambodian-Vietnam border.

A major responsibility for the ground operations is being assumed by South Vietnamese forces. For example, the attacks in several areas, including the Parrot's Beak that I referred to a moment ago, are exclusively South Vietnamese ground operations under South Vietnamese command with the United States providing air and logistical support.

There is one area, however, immediately above Parrot's Beak, where I have concluded that a combined American and South Vietnamese operation is necessary.

Tonight, American and South Vietnamese units will attack the headquarters for the entire Communist military operation in South Vietnam. This key control center has been occupied by the North Vietnamese and Vietcong for 5 years in blatant violation of Cambodia's neutrality.

This is not an invasion of Cambodia. The areas in which these attacks will be launched are completely occupied and controlled by North Vietnamese forces.

Our purpose is not to occupy the areas. Once enemy forces are driven out of these sanctuaries and once their military supplies are destroyed, we will withdraw.

These actions are in no way directed to the security interests of any nation. Any government that chooses to use these actions as a pretext for harming relations with the United States will be doing so on its own responsibility, and on its own initiative, and we will draw the appropriate conclusions.

Now let me give you the reasons for my decision.

A majority of the American people, a majority of you listening to me, are for the withdrawal of our forces from Vietnam. The action I have taken tonight is indispensable for the continuing success of that withdrawal program.

A majority of the American people want to end this war rather than to have it drag on interminably. The action I have taken tonight will serve that purpose.

A majority of the American people want to keep the casualties of our brave men in Vietnam at an absolute minimum. The action I take tonight is essential if we are to accomplish that goal.

We take this action not for the purpose of expanding the war into Cambodia but for the purpose of ending the war in Vietnam and winning the just peace we all desire. We have made - we will continue to make every possible effort to end this war through negotiation at the conference table rather than through more fighting on the battlefield.

Let us look again at the record. We have stopped the bombing of North Vietnam. We have cut air operations by over 20 percent. We have announced withdrawal of over 250,000 of our men. We have offered to withdraw all of our men if they will withdraw theirs. We have offered to negotiate all issues with only one condition - and that is that the future of South Vietnam he determined not by North Vietnam, and not by the United States, but by the people of South Vietnam themselves.

The answer of the enemy, has been intransigence at the conference table, belligerence in Hanoi, massive military aggression in Laos and Cambodia, and stepped-up attacks in South Vietnam, designed to increase American casualties.

This attitude has become intolerable. We will not react to this threat to American lives merely by plaintive diplomatic protests. If we did, the credibility of the United States would be destroyed in every area of the world where only the power of the United States deters aggression.

Tonight, I again warn the North Vietnamese that if they continue to escalate the fighting when the United States is withdrawing its forces, I shall meet my responsibility as Commander in Chief of our Armed Forces to take the action I consider necessary to defend the security of our American men.

The action that I have announced tonight puts the leaders of North Vietnam on notice that we will he patient in working for peace; we will be conciliatory at the conference table, but we will not be humiliated. We will not be defeated. We will not allow American men by the thousands to be killed by an enemy from privileged sanctuaries.

President Richard M. Nixon
Speech given to the nation on Cambodia
April 30th, 1970

"THEY'RE COMING IN...FIX BAYONETS!"

(May 12th - 13th, 1970)

I had been in Vietnam with the 2nd Platoon of Bravo Company, 5th Battalion, 12th Infantry, 199th Light Infantry Brigade since arriving in-country as a scared but optimistic draftee on July 4, 1969.

Bravo Company, along with the other four infantry companies in our battalion, had for nearly six months operated deep in the remote and far-flung boonies around Firebase Libby and Gladys in Long Khanh Province. Daily, we humped through hell-like heat and gushing jungle rains, searching in near vain for the phantom-like little people of the 274th Viet Cong Main Force Regiment and their hardcore brethren from the northern regions that comprised the 33rd NVA Regiment.

Sometimes, when not pounding and trudging through the lush, tropical jungle like mindless green zombies, we would catch the enemy off guard while on ambush, grimly netting a body count of 2 or 3 enemy soldiers killed in action. Sometimes, the Viet Cong and North Vietnamese would catch us off guard...

For several months, we lived in this nightmare epitome of jungle warfare, with young men barely old enough to vote or buy a beer back home in the world, fighting and dying in hundreds of small but deadly firefights and ambushes that never grew beyond the company level.

Most of my tour in Vietnam was like this. Then, the battalion went into Cambodia. Shit.

Tuesday, May 12th, 1970 dawned as another unbelievably hot, sweltering day in Southeast Asia. It ended with an unusually heavy and almost blinding monsoon rainstorm that utterly drenched the already saturated jungle. For the exhausted and apprehensive young infantrymen from Bravo and Charlie Companies of the 5th Battalion, 12th Infantry "Warriors," 199th Infantry Brigade (Sep)(Lt), that day, especially late that evening and into the early morning hours of May 13th, 1970 was anything but ordinary.

Unbeknownst to them, before the sun broke over the horizon ushering in a new day, they would fight in a savage and desperate contest for survival against an equally motivated and battle-tested enemy. This battle would also be the epitome of a classic Infantry battle, with nothing more than 60mm and 81mm mortars supporting both sides for fire support. For the small number of men from the separate, 199th Light Infantry Brigade, the next forty-five days in Cambodia were worse...

Eleven days prior, after spending nearly six months routinely patrolling the rolling hills and triple canopy jungles of Long Khanh and Binh Tuy Provinces in the III Corps Tactical Zone of South Vietnam, the 5-12th Infantry and Delta Battery, 2-40th Artillery, were placed under the temporary, operational control in the 2nd Brigade of the large and flamboyant 1st Cavalry Division (Airmobile).

The Brigade itself was then under the command of Colonel Robert W. Selton, who was hastily given responsibility for the unit in a confused flurry of activity after Brigadier General William Ross Bond was killed in action after a particularly heavy firefight with the 33rd NVA Regiment on April 1st, 1970. (When "Bad Ass" Billy Bond was shot by an NVA soldier after jumping from his command helicopter that day, he became the only general officer killed in combat by direct enemy ground-fire during the entire course of the Vietnam War).

According to one line infantryman in 5-12, "When we heard that we were going to be op-conned to the 1st Cavalry, we thought that it was going to be smooth sailing. No more walking around all over Vietnam for a ride or for an extraction. I mean, the guys in the Cav had hundreds of choppers. We heard rumors from other soldiers that the slicks from the Cav brought in the grunts' rucksacks, complete with coffee and doughnuts almost every morning. Boy, were we in for a big surprise."

Largely unknown to the enlisted men in the ranks, this rapid shift in command and control was done in anticipation for the much-awaited cross-border Cambodian Incursion, which began with both a military and political boom on May 1, 1970.

Combat elements of the 1st Cavalry Division, along with the 11th Armored Cavalry, 25th Infantry Division, 4th Infantry Division and the 3rd Brigade of the 9th Infantry Division would spearhead this massive and long overdue assault.

Beginning in the early morning hours of May 1, 1970 after a literal ground-shaking mission of thirty-six B-52 "Arc-Lights," a World War I vintage, walking artillery barrage that included over 100 artillery pieces quickly followed by 151 tactical air strikes, a Specialist Fourth Class from Charlie Company, 2nd Battalion, 7th Cavalry became the first American trooper to combat assault into the "Fishhook" region of Cambodia.

For those organic and attached units of the 2nd Brigade, 1st Cavalry Division being deployed to Cambodia, specifically into the Fishhook, their primary mission was to neutralize COSVN Headquarters (Communist Office of South Vietnam) and wipe Base Area 351 from the face of the earth.

On May 3rd, after a pleasant but intoxicated two-day stand-down at Camp Frenzell-Jones, the Brigade Main Base for the 199th Infantry Brigade, the entire "Warrior" battalion, along with their cannon-cocking comrades from Delta Battery, 2-40th Artillery, reluctantly made the journey northward from Bien Hoa Airbase towards the infamous "Fishhook" region of the Cambodian/Vietnamese border.

Reid Mendenhall, a combat veteran of Charlie and Echo (Sniper) Companies, 5-12th Infantry describes the wave of emotions that came along when they were told they were going to Cambodia.

"It all started with rumors, like so many others things and nobody could have imagined something quite so bazaar. We all knew that Cambodia was a major problem, but we thought that it was out of reach, until now. We had heard stories about covert units going into Laos and

Cambodia, but now the story was ours. Soon after the rumors began, the word finally came down that we were going. Cambodia. There was some trepidation in our small unit, but we were all together, at least for the moment, and that was of great comfort."

Ironically, when the battalion reached Bien Hoa, they were given their monthly pay. Many soldiers in the battalion wondered, "Just what in the hell are we going to do with this money in Cambodia?"

(When the Warriors of the 5th Battalion, 12th Infantry were given orders for their cross-border adventure, the battalion was commanded by LTC David A. Beckner, who had been in command since October 17th, 1969. Beckner had entered the U.S. Army in 1950 and prior to his joining the 199th LIB, he had served with the Headquarters Section of the United States European Command in Stuttgard, Germany).

Packed aboard noisy and lumbering Air Force C-123 cargo planes at the 8th Aerial Port at Bien Hoa Airbase, the trip north was a solemn one as the young soldiers could sense that they were getting into something big. The mood was ominous. Horseplay and joking around was kept to a minimum as the men were preoccupied with thoughts of home and a foreboding premonition about what was to come. By 1905 that evening, the airlift was completed at Duc Phong, with all companies of the battalion, including A & D Teams from Company M, 75th Infantry (Ranger) on the ground and dispersed to FSB Buttons.

From May 4th to May 11th, the Redcatchers conducted mundane screening operations and short-range patrols along the Vietnamese side of the border at Song Be and Bu Dop. These villages were located in the immediate vicinity of fire support bases Snuffy, Lee and Buttons in northern Tay Ninh Province. Elements of the battalion also guarded the dilapidated earthen airstrip at Bu Gia Map, located just two kilometers from the Cambodian border.

In the meantime, Delta Battery of the 2nd Battalion, 40th Artillery was busy thumping out 105mm howitzer fire on suspected enemy targets in the new but temporary area of operation. (Delta Battery was created on October 27th, 1968 specifically for the 5-12th Infantry. This battery was the final step in having an artillery unit allocated specifically for direct fire support for each of the Brigade's four infantry battalions, thus giving the 199th LIB the flexibly in being the only light and separate infantry Brigade in Vietnam).

Allen Thomas had been with the 2nd Platoon of Bravo Company in the field as an infantryman since July, 1969. He had 63 days to go in-country when the battalion moved to Cambodia.

"My platoon was at the airstrip near Bu Gia Map for what seemed like days on end. It really made me nervous to set up in the same position every night. We didn't fortify and we were spread so thin, I can't remember ever seeing the next position. I do remember the C-130's and other planes coming in on such a short runway. It was also the first time I had seen reversible-pitch props in action. Those great fans literally stopped the airplane in its tracks. Amazing. The water point was near the end of the runway and it got sniped at almost everyday."

John Wensdofer had been with Charlie Company since January 15, 1970. After the two-week Redcatcher jungle school at Camp Frenzell-Jones and less than twenty-four hours after packing his gear, boarding a UH-1 Huey "slick" and joining the company in the field on February 2nd, he was wounded in action.

According to Wensdofer, "We set up a typical night defensive position (NDP). The claymore mines were all out, our weapons were at the ready and then suddenly, out of nowhere, walked up this Viet Cong soldier, right into the kill zone. He hit one of our trip-flares. We blew our claymores and then opened up. The gook and several more behind him also opened up with their AK-47's and hit the M60 gunner next to me. I immediately took over the gun and fired about two-hundred rounds before the firing stopped. When all was quiet, I felt my arm start to burn and noticed blood dripping. A round had hit my left arm during the firefight. I didn't know that I was wounded. In retrospect, I was glad that this happened because I was immediately accepted into the platoon, versus what the other new guys went through. They had to go through an initiation process of sorts and prove themselves over a period of weeks or months. I also started carrying the M60 machine gun at this time. When we started the move to Cambodia, we were sent up north to Song Be for a little over a week. As soon as we got there, the shit really started to hit the fan. They expected the place to be hit and hit hard when we moved in. Where we had been operating in Vietnam, the Viet Cong rarely moved in groups larger than four or five soldiers. Around Song Be and once inside Cambodia, they moved in groups anywhere from 20 to 100."

Bob Pempsell served as an artillery reconnaissance sergeant in Delta Battery, 2-40th Artillery and humped with Delta Company, 5-12th Infantry both in Cambodia and after. During that time, he wrote several meticulous letters home to his sister Judy about what was going on in the days before the unit crossed over the border.

> *4 May 1970*
> *Judy,*
>
> *Well, I just got back from four days in the jungle. It wasn't bad at all, we didn't see anyone, which I was glad. I did a lot of walking though, and saw an eight foot python snake. I kept clear of him. We came upon an old VC base camp and found around 200 hand grenades and some other types of explosives. Guess where parts of the 199th are going? I don't know for sure, but a good possibility is Cambodia being our destination. D Company is supposed to be going tomorrow, and of course, I will be going as well. I'm doing fine as usual and aside from being in Vietnam, I can't complain. Guess that wraps it up for now. Until next time, goodbye. Maybe when you next hear from me, I will be writing from Cambodia.*
>
> *Bob*

Even though things were popping for the advancing U.S. and ARVN forces just a few miles inside the Cambodian border, the Communist forces in South Vietnam did not sit idly by and melt into the surrounding jungle while their Cambodian sanctuaries were being discovered and destroyed. There were several small firefights and rocket/mortar attacks involving the Redcatcher soldiers while in this sector.

On May 8th, after receiving several incoming enemy rounds from the "rocket belt" around FSB Buttons, gun section two of Delta Battery set an unofficial record within the 2-40th Artillery when they countered the Communist barrage with 72, 105mm high-explosive rounds within the span of twelve minutes. During the total fire mission, 356 rounds were sent shrieking downrange. According to a Brigade after action report, "The bunker complex from which the rockets were fired was later to found to be obliterated, with several of the bunkers razed to ground level from direct hits."

(On April 17th, 1970 the 2-40th Artillery fired its 1,000,000 round in support of combat operations since it arrived in Vietnam in December of 1966. The battery also averaged firing 250 rounds per day while there. From May to June, 1970 the 2-40th Artillery had elements of its batteries in three provinces: Long Khanh and Binh Tuy Provinces in Vietnam and Khet Mondul Kiri in Cambodia. It was the only artillery unit in the Republic of Vietnam that had guns giving direct and indirect fire support to infantry units extending the span of the country, from Cambodia to the South China Sea).

While the 1st Platoon of Delta Company was conducting a short recon patrol outside of FSB Buttons on May 9th, they engaged four enemy soldiers moving south on a trail that intersected their line of march. They were immediately taken under fire by the platoon's M16's, M60's and M79's.

Bob Pempsell describes this action in another letter home.

9 May, 1970
George and Judy,

I don't know when you will get this letter. I am writing it out in the jungle and since we can't have things picked up, I'll have to wait until I get back. First off, I am not in Cambodia yet. We landed at Song Be. That's the home of the 1st Cav., 2nd Bde. They are the ones that are in there (Cambodia) now. Hit my fist landing zone (LZ) today, luckily there was no enemy there. We traveled about 1000 meters. While en-route, we came upon two, huge bomb craters that were probably made from 750lb. bombs. Really, you could have fit your house in it. Honestly, it was that big and deep. Now for the war news. After those bomb craters, we came upon a small river. It even had falls and would have made a good picture but I don't have a camera. On either side of the river, we found fresh footprints and two blood trails. After we moved 500 meters, we set up for the night. While we were doing this, 1st platoon, which was 300 meters away, ran into a couple of gooks. When they opened fire, man, was I eating dirt. But they took off, so nothing came of it. One of the guys killed a bamboo viper. It's a very poisonous snake. It was yellowish-green in color, slim and about three feet long. The terrain here is very hilly, with plenty of streams and bamboo. The going is rough and it's a bitch walking in this stuff. Please tell mom that I am 10 miles east of the Cambodian border. It is hot as usual and rain is a regular occurrence at night. There is nothing wrong with me, only ten more months to go. So long for now.

    Bob

With each passing day, the young men from the 199th LIB wondered just when the time would come when they too would cross the border and take part in what was commonly called in GI jargon as, "The Great Souvenir Hunt."

That time came on the morning of May 12th, 1970 when Colonel Carter Clarke, the commanding officer of the 2nd Brigade, 1st Cavalry Division, summoned up his reserves. The Redcatchers were ordered to take up the slack left by the 5th Battalion, 7th Cavalry and quickly re-deploy to a place called LZ Brown after the Cavalry Soldiers had been ordered to assault an enemy basecamp several klicks to the north of the small landing zone.

Landing Zone Brown had been a target hit by the 5-7th Cavalry on the first day of the Incursion. Located approximately three kilometers north of the border and two kilometers west of Route 14-A (one of the only major roads in the region) in a natural clearing that was surrounded by the thick Cambodian jungle, the "hole called LZ Brown" was used as a patrol base by the 5-7th Cavalry for the first week and a half of the Cambodian operation.

Ten kilometers north of LZ Brown was FSB Myron, a firebase that was used by the 2nd Battalion, 12th Cavalry. As with nearly all firebases in Vietnam and Cambodia, they were within mutual supporting distance of one another so as to provide interlocking artillery and logistical support. The two bases however, sat right in the middle of a huge NVA bunker, cache and logistical complex.

(FSB Myron was named in honor of Major Myron Diduryk, a former operations officer in the 2-12ᵗʰ Cavalry. He had been killed in action a month earlier during a firefight in Vietnam. The firebase was located approximately four klicks from "Rock Island East." Named after the sprawling federal arsenal in Illinois, Rock Island East was one of the largest arms and ammunition cache's found by the Americans during the Vietnam War. Three hundred and twenty-six tons of arms and munitions, sitting above ground on wooden pallets and covered in plastic, were found by D/2-12ᵗʰ Cavalry on May 8, 1970).

While other parts of the 5-12ᵗʰ Infantry were heading towards LZ Brown, the snipers from Echo Company were airlifted into FSB Myron by Chinook several days before the rest of the battalion deployed into Cambodia. According to Reid Mendenhall of the Sniper section of Echo Company, "Unlike other firebases in Vietnam, Myron had not been sprayed with defoliation around the perimeter and the jungle extended very close up to the bunker line. While here, we pulled perimeter security and went out on LP/OP (Listening Post and Observation Post) details. We made a few sweeps at different compass headings, but because we were such a small unit, we tried to remain hidden as we had heard that the NVA moved around in battalion strength."

By midday on May 12ᵗʰ, the 5-7ᵗʰ Cavalry had pulled out and left LZ Brown, taking anything of use with them and leaving the small perimeter all but vacant and bare. The "base" now consisted of a pitiful-looking dirt berm, roughly four feet in height, with a single strand of dull and floppy razor wire that surrounded the oval perimeter.

"When we received the word to go into Cambodia," relates John Wensdofer, "Charlie Company flew in by Huey late in the afternoon on May 12." "Earlier in the day, Charlie Company had been right in the middle of a thick bamboo grove back at Song Be, protecting some sort of water purification machinery. When the word finally came down that this was it, we were going over the border, I was thinking that that place was the last place in the world I wanted to go. We were the first ones from the battalion to cross the border. Bravo Company came in later that evening, I believe, by Chinook. For some reason, we dug no foxholes or fighting positions on our side of the perimeter (Charlie Company deployed to the western side while Bravo took up positions on the east). All that separated us from the tree-line was a four foot dirt berm and a flimsy strand of concertina wire."

(The 1st Platoon of Charlie and the Company CP were the first elements of the battalion to combat assault into LZ Brown. They began the lift at 1800 hours and were flaring into the landing zone at 1810. It took six sorties for them to get in. The rest of the company soon followed. Bravo Company did not start the move until 1825. After 17 sorties, all of Bravo was inside by 1930, quite late to be settling into a new and unfamiliar area of operations).

Allen Thomas of the 2nd Platoon of Bravo Company remembers that, "When the Chinooks started ferrying troops out, it was late in the day. Everyone who could, held back and waited. The basecamp, when viewed from the air, looked like a very small, very dirty hole in the middle of the jungle. We built no bunkers nor did we dig any foxholes. I was on the last Chinook to arrive and it was almost dark when I walked over the berm. In the previous couple of months before we made the move into Cambodia, we worked closer and closer towards the border region. Most of our missions took place in the 'Wichita' AO near FSB's Libby and Gladys. We patrolled for either a few days or a week. The jungle there was very, very thick. When we made contact, it was often small, quick and deadly. However, we had a decent routine going there. We would pull an average of 9-10 days in the field and then rotate back to FSB Libby for a much-needed three day rest period. There were 18 bunkers on the perimeter there, just big enough for one company to pull guard."

Terry Braun, an experienced veteran sergeant and squad leader with the 3rd Platoon of Bravo Company gives his detailed account of arriving at LZ Brown that afternoon.

"My squad arrived by Huey's about an hour before sunset. It was a magnificent sight. The afternoon rains were finally beginning to lift and the sun sparkled off the lush, green vegetation surrounding LZ Brown. It almost had the appearance of a well-kept golf course. LZ Brown consisted only of dirt surrounded by a berm. The entire base could have fit inside a football field. It was more oblong than circular and had a smaller berm located within the larger perimeter towards the western half. One single strand of concertina wire surrounded the entire landing zone. It was located about 35 to 40 meters from the berm. There were no bunkers, just mud and dirt pushed up about chest high. Those of us arriving had no idea that the Ho Chi Minh Trail was approximately 100 yards to our west. In Vietnam, we normally worked in platoon sized groups to defend firebases. Now, two Companies were given the task of guarding Brown. Also, as we were leaving the Bien Hoa airport a few days earlier, we were issued M8 bayonets for the first and only time in Vietnam."

In an August 2005 interview with <u>Vietnam</u> magazine, Michael V. Meadows, a combat engineer moving towards LZ Brown with the 31st Combat Engineer Battalion, 20th Engineer Brigade recalls, "All that first day (May 12) we moved toward FSB Brown, where two companies of the 199th Light Infantry Brigade (LIB) had been airlifted in ahead of us. At nightfall, we were just coming within sight of the firebase. There was a swamp separating us, so we remained on the far side overnight in order to find a route over solid ground at first light. We circled our vehicles and pushed up a small berm around our perimeter with the Rome plows. Too tired to dig foxholes, many of us crawled under vehicles or slept behind the tanks and APC's."

By the time complete darkness had crept in, only Bravo and Charlie Companies of the 5-12th Infantry and one 81 mm mortar from Bravo Company's mortar platoon had been airlifted into the perimeter of LZ Brown. All other lifts for the night had been cancelled.

CPT David Ashworth, the S-4 officer for the Warrior battalion during Cambodia states, "When all the lifts were called off going into LZ Brown, only Bravo and Charlie Companies were on the ground. No artillery pieces from Delta Battery made it in that night. There was, however, an advance party of artillerymen that went in and staked out the area for the 105's. The artillery guns were not flown in until the morning of the next day. In the beginning, there was some confusion at battalion about where Bravo and Charlie had been inserted. Actually, they were to be inserted into another area to begin with, but their orders were actually changed in mid-flight. I know this, because I was charged with supplying these units out in the field and I had to ask where they were at the nightly staff meeting."

Because all further lifts were cancelled going into Cambodia, the weary troopers from Headquarters Company, Alpha and Delta Companies and the artillerymen of Delta Battery spent another night on the Vietnamese side of the border, nervously anticipating the next day when they too would cross the border.

For the tense and jumpy troopers of Bravo and Charlie Companies, all that could be done the night of May 12th was to set up the small number of claymore mines and trip-flares that were on hand a few meters beyond the perimeter wire, eat a cold meal of C-rations and huddle silently on the soggy and muddy ground, hoping that nothing major happened until the next day. There was, however, an unexplained, deep feeling of apprehension amongst most there that night that something big was about to happen.

Terry Braun remembers this uneasy premonition.

"That evening, we sensed that something was different. As we positioned our claymore mines by the concertina wire, we listened to the Armed Forces Radio Network, which was something we normally did at basecamp as we were getting ready to pull guard. We learned that the anniversary of Ho Chi Minh's death always brought more activity, and then we heard about the Kent State shootings. We were appalled and saddened by the actions of our National Guard units back home. We set up five soldiers to a position that night, not the usual four. LZ Brown was split in half. On a clock, Charlie Company took the western half from 6 to 9 to 12. Bravo Company took the eastern half from 12 to 3 to 6."

In the surrounding jungle, the change in tempo and activity at LZ Brown during the day did not go unnoticed by the ever-present NVA, who were lurking deep in the confines of the surrounding jungle. As early as mid-afternoon on the twelfth, the 1st Cav's, 2nd Brigade intelligence units began receiving reports from roving aviation and reconnaissance patrols that the NVA had watched the cavalrymen of 5-7 vacate the base. There were even a couple of reports that came in stating that large numbers of NVA had been spotted moving in the direction towards LZ Brown.

Quickly moving to the LZ were hard-core NVA regulars from the 174th North Vietnamese Regiment. Originally formed in North

Vietnam in the early 1960's, the regiment had seen considerable combat in the II and III Corps Tactical Areas of South Vietnam. It was now a part of the 5th Viet Cong Division.

Barely two and a half years before in November of 1967, the 174th NVA Regiment fought against the 173rd Airborne Brigade and the 4th Infantry Division in one of the Vietnam War's largest and costliest battles. The battle was called Dak To and it was fought on a little-known mountainside called Hill 875. By the time the hill was secured by the weary paratroopers on Thanksgiving Day, 1967 the 174th NVA, along with several other supporting Communist units had lost over 1,600 personnel killed in action.

Another enemy unit that was in the 5th Viet Cong Division and that fought alongside the 174th NVA against the 199th LIB in Cambodia was the 275th Main Force Viet Cong Regiment.

The 275th VC and the 199th LIB were no strangers as they had met before. During the horrific fighting of the Tet Offensive of 1968, the 275th Viet Cong Regiment had stalwartly hurled themselves time and time again at the perimeter defenses of Camp Frenzell-Jones in massive human-wave assaults. On the morning of February 1, 1968 over 500 enemy dead from the 275th lay twisted and mangled in BMB's perimeter wire. The dead bodies were subsequently buried in a mass grave near Ho Nai Village.

"We knew when we moved in we could be hit," recalls First Lieutenant Edward Olson, 1st Platoon's leader in Bravo Company, "so, we got set for anything."

Sometime after midnight on May 13th, 1970 the last of the sputtering afternoon monsoon rains slackened and then finally quit. The muggy and humid temperatures immediately returned, along with the serenading noises of millions of tropical insects and mosquitoes. There was also another sinister type of activity that began when the rain ended on that early morning hour.

Unnoticed and slowly inching closer and closer to the dirt berm of LZ Brown were fearless enemy sappers. Clad in dark loincloths with their deliberate movements and sounds concealed by the thin layer of ghostly fog that was emanating from the tree line, the sappers proceeded to disarm or cut the wires on the claymore mines that were pointed outwards, surrounding the American positions.

On Charlie Company's side of the perimeter, John Wensdofer, draped in heavy belts of 7.62mm machine-gun ammunition, peered into the black darkness from behind his M60. Suddenly, less than ten feet to his right, an enemy sapper with his AK-47 at the ready, jumped up on top of the berm and cut loose with a long, sweeping burst of automatic rifle fire.

According to Wensdofer, "The AK fire hit my squad leader, who was a muscular black sergeant with a size 15 jungle boot. The rounds missed me and plowed into the foot of Sp4 Tate who was to my right. Despite being hit, our squad leader then killed the NVA soldier on the berm."

Sp4 William Barry, who was also on that side of the berm, saw the opening shots. "One enemy soldier got on the berm and was shooting into the camp. He was the first to get it."

Malcolm Smith had previously served as a platoon sergeant in the second platoon of Bravo Company from September of 1969 until April of 1970, at which time he was moved to the Company CP. When the battalion moved into Cambodia, he was then on his second of three combat tours in Vietnam. Smith recounts his version of the opening shots.

"Shortly before the fireworks started, I remember making the rounds and speaking briefly with exhausted Bravo Company soldiers. At approximately 0310 hours, the firing suddenly began in the C Company area, followed by silence. Our one mortar fired a flare or two to illuminate the night. After this brief exchange of gunfire, which was immediately followed by the crump of grenades and the deafening crack of claymore mines, there was a brief silence around the firebase that seemed almost surreal."

Smith continues. "Then, all hell broke loose."

Terry Braun, who was just a few meters from SSG Smith remembers, "My second shift of guard duty was to begin at 3:15 AM. At times there seemed to be an internal alarm clock that woke us up just prior to the start of our shift. This was the case for me on May 12th. At 3:00 a.m., I awakened and watched PFC Davis from Tennessee pull his final few minutes of guard. I approached the berm and joined him. His attention was towards Charlie Company's position. He pointed out to me three silhouettes standing on the berm. His first comment was that although those guys must be having a tough time sleeping they should not be silhouetting themselves in the moonlight on top of the berm. We immediately knew then that was not something a GI would do. Seconds later, M16's and M60's started firing and a claymore mine blew. Then the battle started."

The waiting NVA in the tree line cut loose and slammed the small base with intensive, ear-splitting rocket, mortar and .51 caliber machine gun fire. For 15-20 minutes, the North Vietnamese lobbed round after round of 60mm mortar and RPG rounds into the perimeter. A full-scale ground attack by an estimated two reinforced companies from the 3rd Battalion of the 174th NVA Regiment, 5th VC Division, was fanatically launched when the mortars stopped firing.

When the flares went up on John Wensdofer's side of the perimeter, he quickly looked over the berm and to his shock, watched as a large human wave of screaming, pith-helmeted NVA soldiers charged crazily out of the tree line.

"When the flares went up, I could see them running at us and I immediately opened up with the M60. On and on they came. I fired so many rounds, the barrel got white hot. We were firing so furiously, I think that we almost ran out of ammo. I know that I had several belts of 100 rounds linked together. It is hard to remember how much time went by from when the gook got on top of the berm to when I stopped firing the M60. I can remember praying for daylight."

Allen Thomas, who was on the eastern side of the perimeter with Bravo Company states, "The attack was mostly on Charlie Company's side of the perimeter. The fire was the heaviest that many of us had heard and experienced in Vietnam and I remember that the NVA were firing all different sorts of tracer rounds at us. On my side, we emptied some M16 magazines, but not at confirmed targets. After 1LT Olsen announced that we were under attack, everyone was immediately awake and armed. We scurried about finding our extra ammo and frags, just in case. Each squad had an M60 machine gun and each man was required to carry 200 rounds of ammo. If you still had M60 ammo, you needed to get it to the gun. We were in five man positions and had earlier marked off a spot on top of the berm to pull guard duty. The spots had to be equal distance from each other to insure proper coverage, both for defense and for overlapping fire. Someone had come around earlier and announced that we were being probed and to be 100 percent alert. It wasn't long after when the mortars and other explosions started. Red and green tracers were lighting up the sky, explosions were all over the place, both inside and outside, and illumination was going up from either parachute flares or from the smaller, hand-held white star clusters. The NVA also tried to attack the southern perimeter. That action did not last long. I can also remember looking up and locating where the enemy was firing from by seeing where their tracer rounds were coming from.

The firing coming from Charlie Companies side was intense. Our spare ammunition and frag grenades were collected and taken to their side."

John Wensdofer can testify as to how dangerously close the contact was. "At times, the fighting was so close on our side, we crawled up to the top of the berm and either emptied a magazine from our weapons straight down at them or just rolled grenades off the top of the dirt and let them roll down."

One of Wensdofer's M60 bi-pod legs was bent double from the force of an explosion from one of the grenades that he rolled over the berm. Several of the claymore mines that had been placed outside the wire and armed before the attack were useless. The detonation wires had been cut or the mine had been removed altogether by the sappers.

At one point in the firefight, some of the men on the perimeter line could hear the NVA shouting commands and blowing whistles at one another. They were that close.

"At the highlight of the attack, it looked like they were coming in," states Captain Gordon Lee, the Bravo Company commander. "The men fixed bayonets and hoped for the best."

"Tracers were coming in and different colored tracers were going out from the berm. It was actually very colorful. The guys at our position, myself included, began singing 'Sunshine Superman' during the firefight for no apparent reason. A Spooky gunship was called in to support us and we could see the tracers from their .51 caliber gun just missing the back of our aircraft. Fortunately, they were not leading the aircraft enough so they continued to miss," recounts Terry Braun.

Jim Horine, who after serving as a line grunt in Bravo for six months, was humping the battalion RTO at the time of Cambodia. While the firefight was raging around him he remembers, "I was inside the LZ perimeter at the CP with Captain Lee. It was the heaviest firefight that most of us had ever seen. The noise from the machine gun and rifle fire was unbelievable. While on the battalion net, I overheard the pilot of the Spooky gunship (call sign Blind Bat) that was supporting us call back and say that he had expended every single round for his mini-guns on targets of opportunity. As he veered off, there was a long line of green, enemy .51 caliber machine gun rounds trailing him."

A Cobra gunship was also dispatched for close air-support. The Cobra's deadly 2.75 rocket and mini-gun fire was called in so close to the fighting that the men on the berm were actually ordered to pull back a few meters so the gunship could make gun-runs up, down and around the LZ's perimeter.

The fighting had raged on for about an hour and a-half when surprisingly, Bravo got a call on the PRC-25 radio's to, "Fix bayonets! They're coming in!!!"

Terry Braun continues. "Fighting persisted until dawn when the Air Force dropped a series of concussion bombs just west and slightly south of our perimeter. When each of those bombs went off, there was a brief instant magnified by stillness in the air followed by a loud explosion that just seemed to suck the air out all around us. The only thing I can recall that was more impressive was being in the vicinity of B-52s when they dropped their payload and the ground actually shook like an earthquake."

For two and a half hours, despite the withering fire coming from the Warrior M60's, M16's and M79 grenade launchers, the fanatical NVA still kept coming.

The lone 81mm mortar crew from Bravo Company was loading and firing illumination and high-explosive rounds like men possessed. While doing so, they were terribly exposed to the incoming enemy fire.

"We conserved our ammunition the best that we could," remembers 1LT Charles Barnes, the Bravo Company mortar platoon leader. "We fired our last round just prior to 5:45 a.m."

Throughout the firefight, enemy bullets and shrapnel whined and bounced around the mortar-men hanging rounds in the tube. About an hour into the firefight, Sp4 Richard G. Desillier of the mortar platoon was shot and killed as he prepared to hang a round in the tube. Despite the loss, the rest of the crew continued to thump out what rounds were available.

Captain David Thursam, the Charlie Company commander at LZ Brown praises the mortar crew. "The mortar gave my men great supporting fire. They moved their gun around and hit on target every time."

"What amazes me to this day about that fight was how we won," recalls another Warrior that was at LZ Brown. "With all of the fire that we were putting out, an equal amount or more was coming in, including mortars and rockets."

Jim Horine watched from the CP as the mortar crew worked. "They fired and fired and fired. I would bet that they set some kind of record for the amount of both HE and Illumination rounds they were thumping out."

"We had a great medic in Bravo Company," remembers Allen Thomas, "named Doc Uhde. He was a Japanese-American from California. The medics carried an enormous rucksack full of medicine, bandages and salves. To keep from carrying the big rucksack around, he carried a smaller Claymore bag. Doc Uhde made his rounds, as he called it, around the perimeter periodically to see if the men needed anything. During the middle of this firefight, Doc Uhde decided go make his rounds. Here came this little, black-haired Japanese-American carrying a shoulder bag, running with his head down. My friend Tony Reece saw Doc and naturally thought that he was a sapper and drew a bead on him. He could make out his silhouette in the flashes of light. Tony Reece, however, never pulled the trigger. Something was telling him not to do it."

While the men of Bravo and Charlie Companies were fighting for survival at LZ Brown, the remaining troops from the battalion could hear and see the flashes of the firefight unfold from the Vietnamese side of the border.

Rolf Hernandez, a young assistant machine gunner in Alpha Company remembers, "We were on the Vietnamese side of the border that night when the attack on LZ Brown occurred. We could see the huge amount of tracers from the tremendous volume of rifle and machine gun fire going on and we knew then what Cambodia would be like." How prophetic his thoughts were.

Combat engineer Michael V. Meadows of the 31st Combat Engineer Battalion was also watching the battle unfold while on the Vietnamese border. "At around 0300 hours, the sky over FSB Brown was set ablaze with tracers and illumination rounds. Everyone in the convoy awakened and we went to 100 percent alert. The battle raged for nearly three hours. All we could do was watch the fight and cheer for our side. A Cobra gunship appeared over the melee, placing supporting fire around the base perimeter. Whenever an enemy machine-gunner fired a burst at the Cobra, he was instantly destroyed by the return fire. Ribbons of red tracers flowed from the air to the ground, sometimes in strings of red lines broken into dashes as the gun was fired in bursts, and at other times in continuous streams like water from a garden hose. It sounded like the ripping of heavy canvas."

From 0300 to 0545, the opposing sides fought one another for control of LZ Brown. At approximately 0500 hours, much needed support from helicopter gunships, Air Force A-37 jets and fixed wing "Shadow" gunships began pounding the attacking NVA from the air. Incredibly, the NVA in the tree line responded to the air strikes and gun-runs with more .51 caliber machine gun fire.

Luckily, when the 81mm mortar fired its last round at approximately 0545, the fight for LZ Brown finally began to fizzle out. Several minutes later, when the first rays of sunlight began breaking over the horizon, the firing eventually ceased and the surviving NVA melted back into the jungle, dragging away large numbers of their dead and wounded. There was, however, some random but inaccurate sniper fire from the tree-line until approximately 0620 hours.

(There is one point brought about by Allen Thomas about the battle that deserves some investigation. All firebases, in both Cambodia and Vietnam, were created so as to provide mutual fire support for one another. However, this was not the case at LZ Brown during the battle. No artillery support from 1st Cavalry Division units at FSB Myron or elsewhere provided any fire missions in support of the battle. Perhaps this is because there was air support, but that asset did not arrive until over an hour after the battle had already started. Another interesting fact, as found verbatim in the official after action reports of the 1st Cavalry Division, states that the contact on May 12th and the 13th was considered to be a, "light ground probe").

With the daylight the rest of the Warrior battalion was quickly airlifted into the base. The combat engineers from the 31st Combat Engineer Battalion, along with their heavy equipment, bulldozers and Rome plows also arrived at the base after literally creating a new road through the jungle. One of the first tasks for the engineers was to dig a deep trench in which to bury the dead enemy soldiers.

The scene around Brown was one of utter death, carnage and destruction. Only those who have fought in war can understand its pitiful sufferings and tragic ramifications. Fifty-three enemy bodies and pieces of bodies surrounded the small perimeter of Brown. Heavy blood trails were seen from the dirt berm leading back into the jungle. Discarded weapons and equipment were strewn everywhere.

The Redcatchers lost one man killed in action along with eight others from Bravo and Charlie who were wounded. (Those wounded in action were 2LT Robert Mosby, SSG Tim Osborn, SGT Robert O'Sullivan, Sp4 Jerry Moore, Sp4 Harry Dunson, PFC James Wenger, Sp4 Dan Riggleman and Sp4 Earl South).

"At the conclusion of the firefight, some of the Redcatchers started to shout expletives and vulgar comments at the retreating NVA, challenging them to stand and fight some more," recalls Malcolm Smith.

"In the aftermath, I remember carrying the dead to the bulldozed hole and throwing them in. Some of the enemy had been hit multiple times and were bandaged more than once. Some were not complete bodies either."

Terry Braun states, "As the sun rose and daylight crept in, to our amazement the area west of Brown was littered with bodies and body parts. It almost looked like an Easter egg hunt with the eggs poorly hidden. Heavy blood trails led back into the jungle. This was a sight that we rarely saw in Vietnam. Usually after a firefight, the NVA would withdraw and we would see very few bodies. My machine gunner from Chicago, Ken Burmeister, raised an American Flag that he had carried with him on a flag pole and placed it on the berm. That picture appeared with headlines on the front page of the New York Times. Burmeister was later verbally reprimanded for raising an American Flag in Cambodia, but at the time it seemed appropriate."

In front of John Wensdofer's red-hot M60 machine-gun were 12 dead NVA. "They were all wearing regular green and khaki, NVA uniforms with pith helmets. There were six enemy bodies in the wire. I took a 9mm pistol from what looked like a Chinese advisor. His body was lying just inside the wire, close to the bottom of the berm. I know he was not Vietnamese as he was taller than the rest and he looked to be about 180 lbs. or more. He even had body fat on him. In short, the view outside the berm was a sickening site to see. This was the first time I saw the enemy clearly. I didn't really want to kill them, but I had no choice. It was either them or you."

Allen Thomas explains that, "One thing that has always stuck out in my mind is that all of the enemy dead were facing the berm. They didn't die retreating."

Several soldiers also recall seeing what they believed were Chinese advisors among the dead. Another odd find on the enemy KIA's were what was thought to be drugs and vials of liquid speed. (It was not uncommon for enemy KIA's to have marijuana in his possession. Some have also described seeing a "white powder" on some of the enemy at LZ Brown. Since cocaine was not introduced until several years later, perhaps this find could be attributed to LSD or PCP, which were common for the time. Another explanation could be either sulpha powder, such as the type used by American GI's in WWII or a vitamin supplement).

According to Smith, "Some of the dead NVA looked too big to be Vietnamese. Most of the bodies were found with speed in their possession. They had fresh haircuts and their uniforms, as I recall, were light green, khaki and even light blue. I also remember that when we started to police up the enemy weapons, their bayonets were affixed. What really stood out in my mind was the number of discarded bandages, blood trails and drag marks that were going in all directions from Brown. The ground was even soft and mushy in spots. I remember thinking to myself that the battlefield looked as if a blood bank blew up."

"I wasn't in the firefight at LZ Brown that night," states Reid Mendenhall from E/5-12, "but we were choppered in the next day. I took part in the cleanup, checked the enemy dead for maps, documents, weapons and personal effects. I remember the burial which was unlike anything I had experienced previously. But I understood this was for the prevention of disease and the stench, which had already began to permeate the air. The engineers dug a large hole with a CAT and pushed all the bodies into the hole. The dead NVA rolled along like four foot lengths of cordwood. I also remember seeing two NVA prisoners who were blindfolded and being interrogated by ARVN soldiers."

Ihor Dopiwka, who came in with Mendenhall after the battles says, "Most of the NVA were either in or in front of the wire. Even though they had been dead a few hours, the stench from the bodies was nearly unbearable. Many of the dead Communists looked as though they were in their late teens. I also remember that as we were checking the bodies for any intelligence, I found a full-color add depicting a 1970 Chevrolet Impala from LIFE magazine."

Many of the NVA bodies were also found to be booby-trapped. When this was realized, the grunts from 5-12 left the enemy where they lay and let the engineers push them into the ditch. A few of the NVA bodies even exploded when they were moved.

Later that morning, the 2nd Platoon of Bravo Company made a short reconnaissance patrol an hour or so after the battle. Allen Thomas of Bravo explains, "Our platoon was the first out of Brown. We crossed the berm to the north and entered the jungle line. Once inside the jungle, we made a sharp left turn, in single file formation and proceeded to the road that ran north and south. In addition to the enemy's tracks, we saw tracks from some sort of wheel, along with footprints and blood. I believe the wheels were from the Chinese machine guns. From there, we went south until we were in the middle of the battlefield and made another left through the dead and crossed the berm into Brown again. My platoon went back to the east side of the berm to continue pulling security, as we didn't know yet if there would be a counter-attack or not. We also found black communication wire coming out of the jungle and going towards the direction of the berm where Charlie company was. I believe that these wires were being used by a forward observer for fire direction."

In addition to the bodies, blood trails and discarded equipment, 34 individual weapons were collected which included SKS and AK-47 rifles, three light RPD machine guns, four RPG's and two 60mm mortars.

Colonel Robert W. Selton flew into LZ Brown after the battle and made the following remarks. "The casualties were probably more one-sided than was usual in a normally well-planned and executed attack by NVA regulars. Without detracting from the valor or professional expertise of our troops, who were in an unfamiliar place in the dark of night, I would assume that the NVA saw the 1st Cavalry Division withdrawal from LZ Brown and believing the LZ to be abandoned, thought that they could walk in. Surprised at discovering American troops in the berm, they were forced to hastily improvise an attack."

(There is good evidence to support, however, that the NVA knew the base had changed hands during the day and what followed was a well-planned and coordinated attack. The monsoon rains, rapidity of movement, firepower and discipline demonstrated by the troopers from the 199th LIB, however, broke the attack).

By 1600, the 3rd Platoon of Charlie Company was sent outside the northern perimeter of Brown 200-300 meters on a short, cloverleaf mission to look for any NVA stragglers or lookouts. As the pointman and his slack were quietly snaking through the foliage, they quickly engaged three NVA soldiers from a distance of 25 meters who were walking towards them. Two of the enemy soldiers dropped immediately from the short bursts of M16 fire while the third soldier turned around and fled in the same direction he had come from. There was no return fire. The NVA soldiers were described as "wearing gray pants and blue shirts." All were carrying AK-47 rifles.

As for the 5th Battalion, 12th Infantry's first day in Cambodia, Allen Thomas sums up of the thoughts of many that fought there. "In Cambodia, we learned that the enemy stood and fought. The enemy soldiers here were smart and hardcore. Most of us wanted to go back to Vietnam, not stay in Cambodia."

Ch. 2

MY CHET
"GI'S DIE"

(May 14th, 15th & 16th, 1970)

On point. Slowly walking, like a shadow, watching and waiting, eyes stinging with sweat, looking left to right. Right to left. I stop and carefully crouch on one knee. The weight of my rucksack is unbearable and it is so hot. "What is that up ahead?" "Is it a rock or a tree?" "Damn. It looks just like an NVA soldier with an AK-47."

63 days and a wake-up to go. "I've got to keep my shit together." After a couple of minutes, I slowly rise and cautiously, ever so quietly move forward again. Something doesn't seem right. I strain to hear under the weight of the steel pot on my head. "What was that?" "Is there something up ahead?" "What if it is a bunker?" With fear and anticipation, I continue on.

This jungle is the thickest I have seen yet. Nothing compared to Vietnam. I have sores and bamboo cuts on my hands, my body and my feet that won't heal. Sweat continues to pour into my eyes. "I can't take much more of this." "Please God, get me out of here alive."

I cradle my M16 closer and tighter, flipping the safety off just in case. Slowly, deliberately, I continue on to the top of the knoll. I freeze. Up ahead, just a few yards, I see what looks like a small footbridge over a creek. There is a sign up front that looks like it has been written in blood. It reads, "My Chet." GI's Die.

The smell of death still hung strong in the air, mixed with cordite, gun powder, napalm and human waste around Firebase Brown on Thursday, May 14th. As the last of the NVA bodies were being buried, the remaining elements from the 5-12th Infantry (including Delta Battery, 2-40th Artillery) had arrived by helicopter that morning and were already busy shooting azimuths and sending patrols out into the thick jungle.

Charlie Company again killed two more NVA soldiers in a hasty ambush that morning at 0925. Just like the day before, the NVA were wearing gray uniforms. Moving forward a few meters from the contact area, they stumbled upon a small cache that contained 500 lbs. of corn, 600 lbs. of rice, one case of AK-47 ammunition, four cases of 8mm pistol ammunition, one bicycle and 10 uniforms. It was the first enemy cache found for the battalion in Cambodia.

Dave Cook had been with Alpha Company as a rifleman since February. He describes Alpha Company's first day in Cambodia. "On our first day, all the line units from the battalion were sent several klicks from FSB Brown on Reconnaissance in Force (RIF) missions. The terrain and jungle here was absolutely terrible. The jungle foliage, versus back in Vietnam, was much, much thicker. It was almost impenetrable in most places and there was bamboo everywhere. The bamboo grew in super-thick stands well over 8 to 10 feet. The place was also infested by huge, black leeches that sought us out everywhere we went. Combined with these factors with the NVA soldiers we were fighting, it made for a very unhealthy place."

Roger Lowery was assigned as an NCO with the battalion's S-4 supply section and was in charge of re-supplying the line units from the air. He recalls, "When I first flew over Cambodia before the battalion officially moved in, I flew over our soon-to-be Area of Operations with LTC Beckner and COL Selton. Even from high in the air, the jungle below looked clean, green and thick. You could definitely tell that you were no longer in Vietnam as there were no bomb craters or dead spots from Agent Orange. Less than a week later, we flew over the same area again. I had a hard time believing that it was really the same place that I had been before. It was not the same. A B-52 arc-light had gone through the area and it was just devastated. It look as if a giant scoop had come in and ripped the jungle to shreds. What a mess it was."

Bob Pempsell remembers that the jungle was so thick and the heat so oppressive in Cambodia that, "We would stop 15 minutes every hour just to catch our breath. You could not see 10 meters ahead most of the time."

On this day, Alpha Company was busting brush trying to find enemy storage depots and cache sites. Late in the morning on May 14th, they were being shadowed by an OV-10 Bronco plane flown by CPT Patrick W. Kellogg and an artillery spotter from the 184th Aviation Company, 1st Aviation Brigade.

CPT Kellogg was well-known among the company commanders and platoon leaders in 5-12 as he had worked with them before in Vietnam, calling in timely and accurate air and artillery strikes for various units in contact.

As Kellogg was making another pass over Alpha buzzing the treetops looking for anything out of place or suspicious, his plane was suddenly cut in two by a hidden NVA .51 caliber heavy machine gun. Kellogg was killed instantly. (This was possibly the same enemy machine gun that had riddled LZ Brown with bullets during the firefight two nights before. The next day, Sniper Team B from Echo Company, 5-12th Infantry found what was left of the aircraft, along with the log book and confirmed that both Kellogg and the spotter were killed by enemy small arms).

Meanwhile, Delta Company, back in Vietnam, continued to hump around Song Be and FSB Buttons as enemy activity in the region had increased significantly since the beginning of the Incursion.

"I haven't shaved or washed in at least seven days and I don't know when I will be able to do that," wrote Bob Pempsell in a letter home.

"This place is now crawling with gooks. We saw about eight on the trails. The grunts killed an NVA soldier today. An M60 got him and nearly cut him in two. His body is still on the trail and it is booby-trapped with the hopes that the NVA will come back for it. Television doesn't even begin to show how it really is. War is hell."

Out of sight and hidden in the thick jungle underbrush 100-200 meters outside of FSB Brown, the camouflaged men of the Sniper teams from Echo Company engaged in "sneak and peek" type missions, being utilized mainly as early warning devices in the event of another enemy ground attack.

According to Reid Mendenhall, an assistant sniper team leader with over six months time in-country at the time of the Incursion, "Three of Echo's sniper teams went to Cambodia and we operated the same there as we had in Vietnam; independently."

The use of four man sniper teams was a relatively new concept, not only in the 199th, but with most Army units in Vietnam. It was not until April of 1968 when the 9th Infantry Division opened its own organic sniper program and school at Dong Tam that the deadly art of long-range shooting took hold.

The 199th LIB established its on sniper program in late 1968 with volunteers from each of the Brigade's four infantry battalions who attended a week-long school at Camp Frenzell-Jones. As a result, two trained snipers were parceled out to each battalion and used as needed. The 5th Battalion, 12th Infantry's utilization of snipers was a new concept altogether.

Ihor Dopiwka arrived in Vietnam in February of 1970 and upon being assigned to the 199th, he joined the sniper platoon of Echo Company as a replacement RTO while the teams were in for a short break at FSB Libby.

According to Dopiwka, "While I served with them in Vietnam and Cambodia, Echo Company functioned quite loosely, with the snipers doing their thing, the recon platoon doing theirs, the Rat Patrol going in another direction and the 4.2 mortar guys pumping out round from the firebase. The company was effective, however, in performing each of our tasks. LTC Becker really helped the sniper teams out and gave us pretty much what we wanted. He even offered us his personal C&C chopper from time to time. In short, he took care of us and nobody dared to mess with us back in the rear. Perhaps this was because the art of sniping has always been seen as an outcast profession. The line grunts also thought that we were nuts because we went out on missions with five people, which was one or two people shy of a LRRP or Ranger team. In the sniper section of 5-12, two qualified snipers carried National Match M14 rifles with ART-1 3x9x40 adjustable ranging telescopes. They were put into teams with an attached pointman/rifleman, one M60 machine gunner and an M79 grenadier. Depending on the area of operations and enemy activity, two teams of eight men would sometimes go out on missions. We carried as much ammunition as we could handle, with the M60 gunners weighted down with as much as 800 rounds. The teams would then separate in the bush and set up within 100-200 meters of each other, observing trails and engaging targets as far away as 800 meters. Air and artillery would be called in on larger targets of opportunity. In Cambodia, the teams had priority on artillery and gunships and could utilize the powerful assets of 105, 155, and 175mm cannon fire as well as 81 and 4.2mm mortars."

Dopiwka continues. "A typical mission was three days out and three days in. Noise discipline and being super-quiet and stealthy was the key. Once inside Cambodia, we started to go out with regular line units in the battalion. This literally scared the hell out of me. Most guys in the company had been in the green-line prior to joining Echo and knew what a fiasco it was moving with 80 to 100 men in the jungle. The noise we made was unbelievable. It was then that I realized how glad I was to be a member of such a small unit."

May 14th started out on a high note for Bravo Company. Still pumped with adrenaline over the one-sided firefight the night before, the 1st Platoon stumbled upon the battalion's first major find in Cambodia, a fifty-ton rice cache approximately five klicks east of FSB Brown. The PRC-25's immediately came squelching to life with message after message being sent back to the TOC (Tactical Operations Center) at Brown about the find. Within an hour after the call, bulldozers and ACAV's from the 31st Combat Engineer Battalion and the 11th Armored Cavalry Regiment began carving roads through the jungle to backhaul the contents of the cache back to the now bustling firebase.

Malcolm Smith remembers that when the combat engineers and armored cavalrymen arrived to help secure the site, "An Australian news reporter came with them and hung around with us the rest of the day. I don't believe that this reporter had a lot of common sense as he wanted us to get into a contact with the enemy. He even went out ahead with the point element. As we were riding back to Brown on top of the ACAV's, he told us that he was leaving and going to another unit because there was no action here. He should have stuck around for a few hours more."

Early on the morning of Friday, May 15th, despite a few random 60mm mortar rounds that the NVA threw in as a wake-up calling card, the infantrymen of Bravo Company ate yet another quick breakfast of cold C-rations, complained when they slid into their 70 pound rucksacks, locked and loaded their weapons and trudged out into the thick Cambodian bush, confident that they would find another enemy cache as they had the day before. Little did any of them know that for the next 26 hours, they would be involved in the firefight of their lives, even more so than the one at LZ Brown. It was to be a completely opposite contact, with Bravo being outgunned and the prey.

"We went out in platoon file just like we had many times before," explains Malcolm Smith.

"As in Vietnam, each of the three platoons in Bravo would operate in cloverleaf formations approximately 500 yards or so from the Company CP. Rarely did we ever work as a whole company. I learned from my first tour in Vietnam, that an infantry company should never do the same thing twice and set a routine. When on patrol, we would try to dogleg and zig-zag on the azimuth we were following. We would try to mix things up so that we wouldn't invite disaster or walk into an ambush. On some days, we would pack up and move out the first thing instead of eat breakfast and then move. The enemy was always watching and little things like this were good practice as they could save your life and the lives of the other men in the unit. We still did these things in Cambodia, but we quickly learned, specifically on May 15th, that a standard infantry platoon of 25-35 heavily armed men was nothing more that bait for the NVA soldiers fighting us there."

By 1000 hours on May 15th, Bravo, hacking and cursing through the thick undergrowth, had swept approximately three klicks north of Brown when they abruptly bumped into a small band of NVA who were waiting for their approach in one of the small, unnoticeable Montagnard villages that dotted the region between FSB Brown and Myron.

Within seconds, the air was filled with thousands of rounds of AK-47, M16, RPD and M60 machine gun bullets, as both sides tried to gain fire superiority over the other. Within half an hour, the firefight had petered with Bravo taking no serious casualties. However, the outcome for the crew of a UH-1 Huey that had dropped off several news journalists from CBS was much different.

As the chopper was clearing the trees out of the small LZ and starting to gain altitude, it was suddenly and violently blown away with a well-aimed round from an RPG. The grunts below stood in awe with mouths agape, as they watched the explosive round careen out of the tree line and hit the helicopter dead center. What was left of the aircraft shuddered for a second and then dropped from the sky, bursting into flames on its way down.

"I had heard that magnesium and aluminum burns super fast, but I had never seen a chopper shot down before," remembers Terry Braun.

"I would estimate that after only 14 seconds or so, only a grisly looking shell of the slick was left. The two pilots and the door gunners were literally thrown from the craft into the river we were standing beside. As they fell, two of the men were still on fire as they hit the water."

Ken Burmeister and Walter Case (Case was a recent transfer from the 82nd Airborne), both from the 3rd Platoon, immediately jumped into the water and began pulling out the badly injured men from the churning current of the river.

Walter Case recalls that when he reached down to carry out one of the crewmen, "He was charred, nearly burned black. They only part of him that was not scorched was under the armpits. They were in tremendous pain."

Terry Braun adds, "During this entire rescue scenario, we were taking heavy enemy fire. I set up alongside the dirt road between where the chopper crashed and the river where the rescue was taking place. I fired my M203A1 (M16 and M203 Grenade Launcher) weapon to cover the rescuers. During this entire time, the cameraman from CBS was running across the road filming the entire firefight. At the time, I couldn't decide if he was incredibly brave or just plain stupid."

Two of the helicopter crewman died before the dust-off arrived. The other two died of their wounds three days later. The helicopter and crewmen had belonged to the 31st Combat Engineer Battalion, which was supporting the battalion at that time.

By 1400, Bravo Company, with the daily monsoon rains soaking the already exhausted Redcatchers to the bone, had moved out of the contact area, once again jungle pounding through the extremely thick Cambodian countryside. The company was paralleling the same river where the chopper had gone down, two hundred meters or so northeast of FSB Brown.

Because the jungle was so thick, CPT Lee ordered the three platoons of Bravo to deploy into a "Company File," or a long, meandering single file line that when looked upon from the air, resembles a strung-out slinky or snake.

The 2nd Platoon was on point, with the 1st and 3rd Platoons cautiously following. Suddenly, the green column came to an abrupt halt. As the grunts in the line of march lit cigarettes and wondered what the hell was going on up on point, the word started to filter back that the point squad had found something. What they found was spooky.

PRC-25 and 77 radio's began squawking as the RTO's and platoon leaders chatted excitedly back and forth. The grunts strained to listen in on the rushed conversations.

At the very front of the column, the young soldier on point had caught a brief and fleeting glimpse of several enemy soldiers on the other side of a small but fast flowing river that intersected his path. Directly to his front was a small, one-person sway bridge that crossed the river. Off to the side of the bridge was a large wooden sign with a skull and crossbones painted in white. Underneath were the words, written in what looked to be blood, "My Chet," which translates into, "GI's Die." The scene was unbelievable. It looked like it was taken directly from a 1930's Tarzan movie. Even the Kit Carson Scout/Interpreter attached to Bravo was visibly shaken as he was telling people, "Don't go in there."

Despite this find and its psychological effects, the 1st Platoon proceeded to cross the bridge, one soldier at a time. Terry Braun even noticed what looked like crocodiles floating lazily in the water as he crossed over.

According to Allen Thomas, "There were three enemy soldiers that ran out of the basecamp away from the column when we approached. We stood in line for a long time while CPT Lee decided what to do."

Once across, the first elements of Bravo Company found themselves in a very large, very well constructed NVA base camp. Bowls still warm with rice, were found sitting on a picnic table made of bamboo.

When the 2nd and 3rd Platoons finished crossing over, CPT Lee formed the company into a defensive perimeter. From the outset, it was apparent that the basecamp was huge and much too large for one regular line infantry company to handle.

Jim Horine, while carrying the battalion radio in the CP group, remembers that the entire complex smelled new. "We could actually smell the freshly cut wood and bamboo before we saw it. There was also the faint hint of a cooking fire in the breeze. The jungle terrain surrounding the complex was some of the thickest I had ever seen. Because I was in the CP, I heard CPT Lee discuss the possibility of setting up a company sized ambush in some of the surrounding enemy bunkers. It was thought that when we entered the camp, the enemy had fled. We were going to hit them when they came back. It didn't work out that way."

Allen Thomas of the 2nd Platoon states that, "This NVA compound was huge. After we formed a perimeter, each platoon started taking out short reconnaissance patrols to find out just what the extent of the camp was and see if there were any NVA around. I took my squad out on a patrol, the same time that another from the 3rd Platoon went out. We moved in different directions. My squad had gone about 50-60 meters before the danger meter started going off. I knew that something was wrong and that we were being watched. You know it is bad when the hair on the back of your neck literally stands up on end. We found several bunkers, the type we had all seen in Vietnam. They were about four feet deep, three feet wide and about six feet long with two feet of dirt for overhead cover. There were two entrances on opposite sides. I don't know how many bunkers were in there, but there were around 12 to 15 in our immediate area. As we found out, going in there was a big blunder."

The patrol from the 3rd Platoon, which went in the opposite direction of Thomas's, was led by Terry Braun and Arlie "Pete" Spencer.

Braun remembers, "CPT Lee summoned Pete Spencer and me to meet with him. He told us to take a few men and run a short recon patrol about 75 to 100 meters outside of the perimeter we had just set up. He suggested that we stay off any trails and then report back to him when we had gone a sufficient distance. We took along our RTO's, three M60 machine gun teams and a pointman. We tried to stay off the trails, but the thick bamboo bordering the camp kept forcing us back onto the well-beaten paths. When we had gone approximately 75 meters, we came upon a gully that ran next to the river we were paralleling. I stopped and told Pete, 'That's it. 75 meters. Let's go back.' Pete quickly shot back, 'The Captain said 75 to 100 meters.' He then pointed out fresh sandal tracks. You could actually see where the NVA wearing them had hurriedly slid down the embankment into the gully and back up the other side. The gully wasn't that large of a physical feature. It was probably an 8 to 12 foot drop down a very slippery and muddy trail. It ran about 50 to 80 feet to where the trail went back up a hill to the other side and it was probably a little less than that to the hill that surrounded the gully to our right. The river was to our left."

Braun immediately called up CPT Lee and told him the situation and about the tracks. Loaning SGT Spencer his M60 team consisting of Ken Burmeister and Vaughn Bartley, he then told CPT Lee that Spencer was going to probe a little further.

Braun continues. "Walter Case, who just hours ago had helped to rescue the burning crew of the Huey that was shot down, was pleading with the patrol to not enter the gully. He somehow knew that it was a bad omen. I was the seventh one to descend the hill and enter into it. My RTO would have been the eighth and final one, but as I was going down the hill, Pete Spencer was slipping and sliding in the mud trying to get up to the top on the other side. While watching him, I took a quick glance at the area around me. I immediately noticed an enemy bunker pointing in our direction from across the river. I put my hand up to stop the squad from going any further and halted my RTO, who was himself halfway down the trail leading into the gulley. Just at that moment, at approximately 1650, there were two ear-splitting bursts of automatic weapons fire. There is a distinct difference in sound between and M16 rifle and an AK-47. The AK is much louder and makes more of a cracking noise when fired. The M16 looks like a toy and it has a milder popping sound. These bursts of fire were from an M16 and my first thought was that Pete Spencer had crawled up the other side, seen some enemy soldiers and fired at them. As I dove into the mud, I suddenly remembered that we had received clean uniforms a few days ago. Now, I am going to be in these mud soaked fatigues for weeks."

Spencer had made it up to the opposite side of the gully when the firing began. However, it was not from his weapon or from anyone else's in the patrol. An NVA soldier, using an M16, had shot Spencer through the back of the head when he had appeared at the top of the gulley. He never knew what hit him. Seconds before the shots were fired, Spencer turned around to help Ron Scarbrough, who was carrying another M60 machine gun. The rounds that killed Spencer also hit Scarbrough, seriously wounding him in the back and buttocks.

After the two bursts of fire, there was total silence in the gulley for several heart-pounding minutes. The men at the bottom of the ditch were sure that the enemy were fleeing as they had done many times prior to artillery fire crashing after them.

That was not the case on May 15th. Within minutes, the 1st and 2nd Platoons at the company perimeter, 100 yards or so back from the 3rd Platoon, began receiving heavy and accurate AK-47, RPD and RPG fire at a distance of less than 75 meters. At least nine grunts were wounded in the opening fusillade. Bravo Company was strung out and under heavy enemy fire. There was nothing for the men to do but pray and spray the jungle in front of them with their weapons on full-automatic. Serpent A1, the O-1 observation plane buzzing over the contact was immediately driven off by a string of accurate ground to air machine gun fire.

Back in the gully where Braun, Case, Burmeister, Scarbrough, Walker and the others were pinned down, the NVA began closing in around them. Instead of running away, the NVA had formed a perfect U-shaped ambush, completely covering the gully where the eight men of the 3rd Platoon were. The river was on their left and the NVA were to their front and right. They opened up on the Redcatchers with a horrendous barrage of automatic weapons fire.

Terry Braun continues. "After Spencer was shot, I grabbed Scarborough and helped him to get back up to the other side of the gully where a medic immediately began first aid. Scarborough then told me that he had caught a brief glimpse of eight or nine gooks on top when one of them hit Spencer. It was then that it sunk in that Pete was dead. He was a good friend of mine. He had been in-country a little longer than me and I had thought of him as a mentor. He spoke about his wife often. He was only 21 and didn't even shave yet. While

I was assisting with Scarbrough, my machine gunner, Ken Burmeister, dropped his own M60 and immediately ran towards where Spencer was lying and started administering mouth to mouth. Covered in Spencer's blood, Burmeister did not know that he was already dead. Because the fire was too heavy for Burmeister to make it back to the gully, he did the only thing that he could do. He played dead and placed his right hand and his .45 pistol under his head. He lay there for over four hours, covered with the blood of Spencer. During that time, the NVA actually crawled up to him, kicked him, took a ring off his finger and left him for dead. He told me later that after several minutes, the hand underneath his head fell asleep and he could not have fired his .45 if he had the chance. Instead, he began praying to God to get him out if this mess."

Within the span of 15 minutes, two out of the three M60 machine guns from the 3rd Platoon were out of commission and inoperable, thus leaving the men trapped in the gully without any heavy fire support, save for their M16's and M79 grenade launchers. Incredibly, the last remaining M60, which belonged to Sp4 Fletcher Walker, took an AK-47 round straight up through the barrel, rendering it inoperable. (The other two M60's had belonged to Burmeister and Scarborough. Burmeister was playing dead and Scarbrough's was lost when he was wounded).

In the enemy basecamp, grunts from the 1st and 2nd Platoons were being shot at and the jungle was so thick around them, their return fire was doing little to stem the building crescendo of AK-47, SKS, machine gun and RPG fire from the hidden enemy.

Allen Thomas recalls getting into one of the enemy's bunkers, but it was too small for several soldiers to maneuver in, so he crawled back out and laid behind what little overhead cover he could find.

Back at the Company CP, air strike after air strike was being called in around Bravo's beleaguered line. "Napalm was being dropped danger close around us and it was not doing a thing to the vegetation. Guys with M79's were shooting them, only to have the rounds bounce around after hitting the trees and stands of bamboo. The NVA's incoming fire was so hot we couldn't even stick our heads up and the CP group was in one of the enemy's bunkers," relates Jim Horine.

At FSB Brown, the remaining infantrymen and artillerymen from Delta Battery, 2-40th Artillery heard the ferocious sounds of the battle and watched as F-4 Phantom's and F-100 Super Saber's dropped their bombs for hours, increasing their feelings of consternation and dread for their fellow Redcatchers.

The jets would come screaming in to make their bomb runs so low and so fast, that the soldiers on Brown watched in amazement as the silver canisters of napalm and 500 lbs. bombs floated lazily over their heads before exploding with deafening roars and concussions.

With all of the jets, spotter planes, helicopters, sights, sounds, explosions and chaos going on around them, the sweating artillerymen from Delta, clad in their heavy flak jackets and steel pots, continued to slam round after round of 105mm rife downrange at the enemy surrounding Bravo.

"The infantry and artillery worked closer together in Cambodia than I'd ever seen before," states 1LT Stan Hogue, the executive officer for Delta Battery during the Incursion.

"The most remarkable thing about Delta Battery was that almost all of the men were fresh troops when it came to actual fighting. We had been stable at FSB's Libby and Gladys for nearly six months and had seen very little action."

By the end of May 15th, the battery had fired in excess of 1300 rounds in support of Bravo Company. To fire this many rounds, the men in each gun section had to work almost non-stop around the clock for the entire day.

For the 3rd Platoon back at the complex, the situation was deteriorating by the minute. As late afternoon approached, ammunition was scarce and nearly gone. Led by Malcolm Smith, he and three other courageous soldiers braved the incessant enemy fire and threw what extra bandoleers and bags of M16 and M79 ammo they could spare to the trapped men in the gulley, all the while being followed by enemy bullets and RPG's.

Other soldiers from the 2nd Platoon crawled forward to lay down suppressing fire. M60 gunner Bob Ward and his assistant gunner, Roger Canter, bravely crawled forward and laid down a steady, withering fire on the NVA overlooking the 3rd Platoon in the gulley. These actions by Smith, Ward and Canter possibly saved the lives of those trapped men there.

CPT Lee placed SSG Ron Orem, the 2nd Platoon Sergeant and SSG Malcom Smith in charge of evacuating the B Company wounded. Medevacs, with hoists and jungle penetrators at the ready, flared over open spots in the jungle above Bravo hoping to lift some of the seriously wounded out, only to be driven off by heavy and accurate ground fire. For over two hours, the heroic pilots made rescue attempt after rescue attempt. Fortunately, there was a brief respite in the fighting at 2030 hours when six of the seriously wounded did make it out before all the other lifts were cancelled. SSG Orem had exposed himself to the heavy enemy fire by placing a strobe light into his helmet and holding it over his head, thereby guidling pilot over Bravo's position. These wounded got out because of Orem's courageous actions. The rest of the wounded would have to hunker down and wait. Many of these men did without morphine, bandages or other sorts of aid, stoically passing it on to those soldiers that needed it more.

(Those soldiers who got out on the afternoon of the 15th were; Sp4 Vaughn Bartley, Sp4 Fletcher Walker, PFC James Blattel, SSG Bernard McMann, Sp4 Ron Scarbrough and Sp4 Tom Leckinger).

Throughout the day, the platoon and squad leaders of 3rd Platoon were desperately trying to make contact with CPT Lee at the CP. However, the NVA were jamming the radio signals, thus complicating the already tedious situation further.

It didn't take long for some of the men of Bravo Company to realize that the NVA could have overrun them at any time, specifically the grunts trapped in the gully. The enemy was shooting down on them from a ledge above and there was virtually nothing, save for a log a little over eight feet in length to take cover behind.

Terry Braun was trying in vain to reach CPT Lee and inform him of his wounded and low ammunition. "When I could not get through, I started screaming to where the other platoons were located, pleading with them to bring a sixty up. Then, Vaughn Bartley, hearing my screams, did something that logic couldn't define. One of the things that I learned in battle was that the difference between a hero and a coward was an extremely thin line. Whatever decision Bartley made was done in a split second with no regard for his own welfare or consequences. With his M203 on full-auto, he fired a long burst, dove for cover and then came back up firing again. Then, without any personal regard for his own safety, he threw down his weapon and ran to where Bergie had dropped his M60 halfway up the hill before helping Pete Spencer."

Bartley didn't make it. Just as he was within a few feet of the gun, a North Vietnamese soldier came out of nowhere and took it, leaving Bartley stunned and weaponless. He immediately hit the dirt and tried to find cover. There was none to be found.

The NVA immediately opened up on him, shooting off his trigger finger and then the middle finger off his right hand. Bartley then took a round in the jaw, completely shattering it from the ear to the bottom of his mouth. The men in the gully thought that he too was killed in action.

After minutes that seemed like an eternity, Terry Braun finally made his way back to the company CP during a brief lull in the firing. He informed the CO of the 3rd Platoon's situation.

During this lull, Jim Horine, in the CP, was calling in artillery and air strikes on the battalion net. He was also talking to COL Robert W. Selton, the commanding officer of the 199th Infantry Brigade. "I was talking to the Colonel and informing him of the situation," says Horine. "Because we knew the NVA were listening, I had to be careful with what I said. We knew that we needed some tanks and heavy stuff to get us out, so I told Selton that we needed 'Clinkety-Clanks.' He understood immediately and signed off with a 'Roger, hang in there."

In the meantime, the artillery fire was called in at less than 50 yards in some places along the perimeter. After the barrage ceased, Cobra gunships from the 1st Cavalry's, "Blue Max" squadron rolled in with 2.75 inch rockets and miniguns blazing. For the next ten minutes or so, they pulverized the earth in front of them.

Just as the Cobra's made their final gun-run before going back to FSB Brown to refuel and reload, two miraculous events happened.

With screams and shouts coming from men in the gully to prepare for another air strike within feet of their location, Vaughn Bartley, who despite being shot in the hand, arm and face, suddenly jumped up from where he had been lying for the past two hours and made a run for the gully.

"Even the NVA were surprised. However, he was not fast enough. They cut loose with a storm of machine gun fire and he was again hit, this time with a deep and searing AK-47 wound to his left leg. Then, RPG's and grenades followed," remembers Terry Braun.

Amazingly, Vaughn was able to make it back to the 3rd Platoon where he was immediately pulled into the tiny perimeter. The soldiers there packed his wounds with mud and strips of cloth to keep him from bleeding to death.

Shortly after Vaughn's miraculous escape, Burmeister, who had tried to rescue Pete Spencer in the opening shots of the ambush, came bursting out of the jungle. With a hail of bullets trailing him, he dove into the gully with the rest of the shocked survivors, covered in blood, but not his own. He had been playing dead beside Spencer's body for nearly five hours.

(After the battle, when Bravo finally made it back to FSB Brown, Allen Thomas watched emotionally as Burmeister grinned from ear to ear and jumped up and down, shouting, "I'm alive, I'm alive, Thank God I'm alive!!!).

With the last remaining Americans accounted for, the Cobra's and artillery once again went to work. The survivors from the 3rd Platoon then retreated back to the relative safety of the company perimeter.

As darkness crept in, the remaining men of Bravo Company were emotionally and physically drained. The harsh reality of what they had just been through and that they would have to spend the night, tightly huddled in a small, company perimeter in the middle of a sprawling NVA basecamp was stoically accepted.

The NVA probed up and down the line all night trying to find a weak spot. The firing would suddenly start up again, reach a crescendo after a few minutes and then die back down. The darkness was brilliantly punctuated by thousands of red, green and white tracers from both the Americans and Vietnamese as grenade and RPG blasts showered shrapnel around them.

Bob Kenna, a grunt in the second platoon remembers that the NVA shouted insults and taunts in broken English at the Redcatchers during lulls in the shooting. Surprisingly, the North Vietnamese never followed through with a major ground attack. Had they done so, the results for Bravo may have been disastrous.

When the first rays of sunlight came streaking over the Cambodian horizon on May 16th, the Redcatchers finally started to breathe a sigh of relief. They were not, however, out of it yet. By mid-morning, the NVA were again probing Bravo's shrinking perimeter.

At 1015, another lethal and continuous hail of AK-47, SKS and RPG fire swept through the American line, pinning their heads down until finally tapering off at around 1230. The firefight burst open again at 1320 when a watchful grunt in the 1st Platoon killed an NVA soldier who was slithering up to shoot an RPG rocket into the perimeter at less than 50 meters.

An overland attempt by Bravo Company of the 1st Cav's, 2nd Battalion, 12th Infantry was made earlier that morning to reach the survivors of Bravo Company at the basecamp. After two hours of hacking and tromping through the bush, they were themselves ambushed a couple of klicks from FSB Brown and suffered one KIA and seven WIA right from the start. Bravo 2-12 immediately pulled back. It was up to the Sheridan's and ACAV's from I Troop, 3rd Squadron, 11th Armored Cavalry to get them out.

Leaving at approximately 0900 that morning, the armored force, led by a Rome Plow from the 61st Land Clearing Group, had to carve a new road out of the jungle to reach Bravo and the enemy basecamp. All day, the Redcatchers could hear the straining and whining of the diesel engines of the vehicles inching towards them. Occasionally, the motors were drowned out by the chattering and boom of machine gun and cannon fire as the tankers re-conned the area in front of them by fire.

Finally, at approximately 1630, the Rome Plow in the lead broke through into the basecamp and Bravo's perimeter. It was immediately followed by Sheridan light tanks and ACAV's that were cutting loose at the bamboo buildings and bunkers with its huge 152mm main gun and M60 machine guns. The flechette rounds from the 152mm literally cut the structures to pieces, much to the enjoyment of the now rescued infantrymen.

As soon as the armored column entered the company perimeter, the wounded were immediately piled aboard the vehicles while the rest of the grunts wasted no time in policing up the battlefield of any discarded weapons, ammunition and grenades. (Eight NVA bodies were counted, along with several blood trails leading back further into the basecamp).

Several of the men from 3rd Platoon, including the platoon leader, raced back into the gully to retrieve the body of Pete Spencer. They made it back just as the column was leaving the area.

The survivors of Bravo, happy to be alive and more than happy to be going back to FSB Brown, did not know that their 26 hour ordeal was not yet over. As a final insult, the NVA had one more trick up their sleeve.

Because the armored column was leaving on the same road that it had used to enter the enemy complex, North Vietnamese soldiers from the basecamp sprinted ahead and ambushed the armored force as it clanked on approximately one klick from where the link up had occurred.

As the unsuspecting column reached the point where the Huey had been shot down on May 14th, the lead Sheridan tank in the column momentarily became bogged down in a rut. According to Bob Kenna, this is when the NVA cut loose with a literal hailstorm of RPG fire. The swoosh and ka-boom of 10-15 of the cone-shaped missiles hitting the aluminum hulls of the armored vehicles completely caught the tankers and infantrymen off guard.

It was utter chaos. AK-47 and heavy machine gun fire stitched the force from beginning to end, back and forth. Men who had survived the last two days without a scratch were seriously wounded in the short-lived firefight.

Roger Canter, who had bravely crawled forward from his secure location to place heavy suppressing fire on the NVA the day before was blown off a Sheridan Tank with severe wounds from RPG shrapnel and AK-47 rounds. Rich Guenther, a squad leader in the 1st squad immediately ran to Canter's aid. As Guenther was trying to patch Canter's serious wounds and stop the bleeding, he too was hit in the thigh by an AK-47 round.

The ambush did not last ten minutes, but the damage was done, however. At least eight more Redcatchers were wounded, along with an equal number of tankers and cavalrymen from the 11th ACR. Now speeding through the crude jungle road at full-throttle, what was left of Bravo and the reaction force made it back to FSB Brown just as the sun was setting on May 16th.

Allen Thomas, who survived the two day ordeal recalls a puzzling and brief exchange between CPT Gordon Lee of Bravo and the battalion commander, LTC Beckner, when Bravo Company finally made it back to Brown.

"When we finally limped back to Brown on the evening of May 16th, LTC Beckner was there to meet us at the berm. CPT Lee marched directly up to him where they exchanged salutes. Lee then said, 'Sir, I would like to report that Bravo Company is combat ineffective,' to which LTC Beckner replied, 'Bring the boys on in Captain, you guys can pull bunker guard.'

Thomas continues.

"I will never forget that exchange since I witnessed it first-hand. Even though the company had lost a great number of men and my platoon was down to only 13 personnel, we were by no means combat ineffective."

The next day, at a company formation at the battalion tactical operations center, the reality of what had happened over the last two days finally began to sink in while the numbness of combat wore off. More than 25 men were no longer in the ranks, many on their way to hospitals in Japan or the United States with serious, disfiguring wounds. The survivors, many suffering from fevers and symptoms of dysentery, stood out in the daily monsoon rain and wondered if things could get any worse.

(After the ambush and firefight of May 15th and 16th, the 3rd Platoon of Bravo Company, because of the high number of WIA's, nearly ceased to exist as a fighting unit. The last week in May, the platoon was sent back to Song Be for a week of rest, refitting and reorganization. It returned to Cambodia, heavy with new replacements, ready for duty the first week in June. Those wounded but who were not airlifted out on May 15th or wounded in the armor ambush on May 16th were Sp4 Michael Reley, Sp4 John Bowden, SGT Samuel Joiner, SGT Kenneth Grimes, SGT Richard Guenther, Sp4 Paul Luna, PFC Duane Schmidt, PFC Wilson Boston, PFC Albert Davis, Sp4 William Cochran, PFC John Green, Sp4 Dennis McCoslin, PFC Robert Swapp, Sp4 Roger Canter, Sp4 Howard Ueda,

Sp4 Ken Huizinga, Sp4 John Quillen, 2LT Edward Olsen, PFC William Easterling, Sp4 Samuel T. Carroll, PFC Albert Davis, PFC John Green, PFC Orin Bradshaw and PFC Lewis Ward).

For the next week, the 1st and 2nd Platoons were on perimeter guard at FSB Brown. Several of the company members pulled LP/OP missions with the snipers from Echo company during this time.

On May 19th, a rather strange incident occurred that tragically claimed the life of 21 year old Robert J. Urbassik.

Ihor Dopiwka remembers Robert Urbassik as, "A super-nice guy. He never took part in the partying and carousing that the other guys did. He was always quiet and kept to himself, but he would help you out in anyway possible. I don't believe that I ever heard him utter a bad word and he was always writing letters home in his spare time."

According to some of the men from Echo that were on LP/OP that night with Urbassik, they set up their tiny perimeter in thick and hilly terrain overlooking an open draw some 150-200 meters northeast of FSB Brown. It was to be a typical three-day mission for the team. As the five men made their way to the overnight position, the rain was beating down on them in torrents and they were having trouble maintaining contact with the TOC at Brown.

At 2010 hours, the snipers called the TOC and reported movement to their north, east and west. The team held their fire and waited, hoping that the movement would pass them by.

As wicked bolts of jagged lighting flashed and danced across the sky, they suddenly heard the hiss and buzz of projectiles ripping past them. Strangely, however, they did not hear the report of any enemy rifles or machine guns.

As the men lay in the prone position waiting to return fire on confirmed targets, Urbassik slumped over with a small wound in his chest. He had been leaning up against a tree, quietly trying to make contact with the TOC back at Brown. As the medic fought to keep him alive and conscious, Urbassik died in the pouring rain from a punctured lung. There was no exit wound. The men there with him believed that he was hit by a dart or an arrow from some sort of crossbow.

The sniper team leader then began calling in artillery, 81mm and 4.2 mortar fire in a three-hundred, sixty degree radius around the team's position. At 2125, commo with the snipers was lost.

Ihor Dopiwka remembers that his sniper team had just come in from the field and had stayed for nearly three days in the very position that Urbassik's team had just inserted in just two hours before.

"It was raining like an absolute bitch and I was lying down on my cot when somebody stuck their head in the door and mentioned that one of the sniper teams was in trouble and the rest of us were going to go back out and bring them in. I jumped off my cot and felt the water go over my ankles. FSB Brown was, after-all, in a dry lakebed. When we formed outside, I remember CPT Lodoen told us to take our poncho's off and leave them here. I thought that this was pretty odd because it was raining so hard. Then, we jumped on APC's and went tearing through the jungle. The rain beat down harder the further we went. It was so dark that night. The only source of light we had was from the bolts of lighting that briefly illuminated our way every few seconds. It was comparable to going through a tunnel with your lights off. The sniper team heard us coming and when we got close enough, the team leader turned on his strobe light and guided us into his position. We quickly loaded the other team members and Urbassik's body and got out of there."

The next day, a small detail from Echo went back out to the site of the LP/OP to inspect the area. Oddly enough, there was not faintest sign of any splintered trees or crushed and splintered vegetation to indicate that anyone was shooting at them. Had the NVA ambushed them, there would surely have been bullet casings, foot prints or shrapnel present.

The only scenario that seems logical to explain Urbassik's death, both now and then, was that he had been hit in the chest with either a dart or an arrow, most likely from a weapon like a Montagnard cross-bow. Most of the men from the battalion had seen these being carried by some of the "yards" in the Montagnard villages that dotted the region between FSB's Brown and Myron. Although unexplained, all the men that served there know that Cambodia was a strange and dangerous place.

Ch. 3

FIND, TAKE AND DESTROY "THE GREAT SOUVENIR HUNT"

(May 19 - June 10, 1970)

"Ripper Bravo 6, Ripper Bravo 6, this is One Alpha Lima 6, Priority, over."

"One Alpha Lima Six , this is , Ripper Bravo 6. Go ahead, over."

"Roger Ripper Bravo 6. We've found something, over."

"One Alpha Lima 6, copy. What is it, over?"

"Ripper Bravo 6, it's absolutely huge. There's rice everywhere."

Since arriving in Cambodia hours after the firefight at FSB Brown ended on the early morning hours of May 13th, CPT Michael Hess's Alpha Company of the 5th Battalion, 12th Infantry had been beating the bush up and down the hilly and imposing terrain in between FSB Brown and Myron for the past week and a half.

The company had made some sort of contact with the NVA nearly everyday, but it was nothing to the extent of the battle at LZ Brown and Bravo Company's two day firefight on the 15th and 16th.

David Cook describes what it was like for the men of Alpha Company at that time. "We had contact after contact, some lasting seconds, some lasting a couple of hours. There were signs of the enemy everywhere we went. At night, we could hear enemy movement all around us. One night, we even heard heavy trucks and what we thought were Russian or

45

Chinese tanks moving to our front. Artillery fire was called in around the clock near our position. We never, and I mean never heard this type of stuff in Vietnam."

Dick Rose had already served as a platoon leader in the 2nd Platoon of Alpha for the past six months before the push into Cambodia. "I had just gotten out of the field and turned my platoon over to Tim Jorgenson and then went on R&R the first week that the battalion was in Cambodia. While in Australia, I was thanking my lucky stars that I wasn't there and no longer in the field. After R&R, I went back to FSB Libby, which was all but deserted. Three days later, I got the call from the company commander to come on over. The 1st Platoon's leader had been wounded and battalion wanted me to take over that unit. This was just after the firefight at LZ Brown. I can tell you, it was a very tense time."

By the summer of 1970, every minute of everyday was tense, stressful and deadly for the young 19, 20 and 21 year old draftees and near-extinct volunteers as they trudged through the wilds of Vietnam and Cambodia.

While their rather fortunate peers back home in "the World" protested and marched in mass gatherings across the country, burned their draft cards and portions of college campuses, the courageous and thankless American Citizen Soldier "took two salt tablets and drove on," despite an all time low in public support and opinion for the war.

At this time, the very nature of the war was changing. President Nixon's "Peace With Honor" and Vietnamization programs were in full swing. Several American units had already left Vietnam for good and redeployed back to the United States, namely two brigades from the 9th Infantry Division, the entire 1st Infantry Division, the 3rd Marine Division and one brigade from the 82nd Airborne Division. Several more combat units followed their lead by the end of the year. The Vietnam death machine was finally beginning to slow down. None of the soldiers still in Vietnam by mid-1970 wanted to be one of the last one's to die and their mood reflected that sentiment.

On May 4th, 1970, the Ohio National Guard opened fire on an anti-war demonstration protesting President Richard Nixon's decision to go into Cambodia at Kent State University. Tragically, four young students were shot and killed while taking part in the march. During the first week in May of 1970, 284 American soldiers were killed in action in Southeast Asia, 58 of which were in Cambodia.

In slang terms, the Cambodian Incursion came to be called, "The Great Souvenir Hunt" by those American GI's that fought there. This was because of the sheer amount of war trophies, weapons and equipment that was uncovered during the two month "ad" venture. Soldiers from all wars, past and present, have always eagerly searched for enemy trophies and sent home the spoils of war.

One of the most sought after and prized pieces of enemy equipment to find and send home was the Chinese or Russian version of the SKS semi-automatic rifle. Developed in 1945 at the end of World War II, the rifle was self-loading and held 10 rounds of 7.62x39 ammunition. A very accurate and reliable weapon, it also featured a foldable spike or blade bayonet that folded under the barrel when not in use. It was chambered to fire the same round as the world-famous and fully-automatic AK-47.

Because the SKS was semi-automatic, it was legal for the soldier who captured the piece to process the rifle and send it home as a trophy, versus the AK-47 and other automatic weapons that were illegal to ship out of the war zone.

For their two-month stay in Cambodia, most, if not all of the members of 5-12 sent an SKS rifle back home as a trophy. There were, however, a few unfortunate men that were denied this privilege.

Many of the weapons were taken or stolen when the owner was preparing to leave Vietnam by various Air Force or rear-echelon personnel. These lowly non-combatants would tell the ready-to-go-home grunt that there was something wrong or missing with the paperwork or that the policies on taking home a rifle had been changed and it was now illegal to take the weapon out of the country. As a result, these "Saigon Warriors" were able to take the SKS home and thrill their families and friends with heroic and adventuresome tales of derring-do.

Bob Pempsell remembers, "When caches were found, specifically those containing SKS rifles, the helicopter pilots from various aviation units would go nuts and come in over the company or battalion radio net and offer to buy them then and there. It was crazy how people acted."

The discovery of 24, Chinese SKS rifles on May 17th near Giong To by Alpha Company, would in turn contribute to some of the largest enemy cache sites found and destroyed during the entire Cambodian Incursion.

By 0930 on the 17th, Rolf Hernandez, an young assistant M60 gunner in the 3rd Platoon, was standing near the end of a platoon file when he noticed, just barely visible out of the corner of his eye, something blue protruding from the dark and wet jungle floor.

After checking the area for booby-traps, Hernandez saw that it was a tarp covering some suspicious looking material. Quickly pulling the tarp aside, his eyes widened and then became as large as saucers when he saw that it was covering a wooden crate full of brand-new Chinese SKS rifles.

After dividing them up between the platoon, the model and serial numbers of each weapon were written down and recorded, then sent back to the rear area at Camp Frenzell-Jones for storage.

Alpha was now in an area that looked as if it could contain more hidden cache sites and CPT Hess went ahead and ordered the company to prepare hasty night defensive positions. The decision, despite the fact that the NVA knew that they were there with all the noise and activity around them, proved to be a fruitful one.

Early the next day on May 18th, Hess ordered each of his three platoons to split up and perform cloverleaf patrols several hundred meters from the Company CP. At 1000 hours, the 1st Platoon, let by 1LT Dick Rose, was cautiously patrolling down a ravine covered in thick bamboo and wait-a-minute vines when the pointman, PFC Timothy Dolan, halted the column.

As Rose carefully walked to the front of the point element, the tension among the grunts rising each second, PFC Dolan turned and looked at 1LT Rose, motioning for him to scan the jungle ahead.

"I looked and saw a camouflaged hootch just a few meters to my front. We looked closer, our eyes slowly scanning the terrain, when we noticed three more similar structures in the immediate vicinity. We crept closer and discovered that the hootches were well-built and contained several sacks of rice that were painstakingly stacked up on bamboo pallets. They were carefully covered in cheap, green plastic. The sacks of rice looked as though they weighed over two-hundred pounds or more a piece. Most of the sacks were in good shape. There were huge, hairy rats scurrying and crawling over each pile. I remember one of the rats raced up the pant leg of one soldier standing nearby. His trousers were not bloused and the rat raced halfway up to his crotch before he gingerly stopped it from going all the way up."

After a complete check of the four hootches, it was estimated that over 120 tons of rice were uncovered, including other significant finds such as truck parts, universal joints, drive shafts and diesel fuel. Alpha's diligence and professionalism had paid off big-time.

The next day, May 19th, CPT Hess sent out more reconnaissance patrols several meters beyond the four hootches, hoping to find other hidden supplies. Several more enemy caches were discovered throughout the day, the contents of each find being more than was discovered on the 18th.

The company had found a huge but deserted basecamp and supply depot. David Cook states, "We followed a large road and it was so wide and so well kept, it looked like our own engineers had made it. At the end of the road was a storybook village. The buildings were laid out in neat, precise rows that were hidden from aircraft by thick, triple canopy jungle that was tied together. Everything was made of bamboo. There were even company streets made of bamboo, split down the middle and stacked close together. Thankfully, the NVA did not put up any resistance when we approached. They just let us walk in and take it."

At days end, another 500,000 pounds of rice and other enemy material were accounted for.

The battalion's Echo Company Recon, operating alongside Alpha in this sector, also found mammoth stockpiles of enemy supplies and munitions.

1LT Wayne Otte, the recon platoon leader states, "We had been out for seven days and were ground reconning through the jungle towards Brown. We hadn't seen any trails or signs of enemy activity since our first day out. As the pointman was cutting his way through the thick bamboo, he saw a small hootch in a heavily vegetated area. We set up security and checked the immediate vicinity for booby-traps before entering the hootch. A search of the structure produced a large, underground storage area containing a weapons and ammo cache."

The hootch contained a literal treasure trove of enemy material. Three-hundred, twenty four B41 rocket rounds and boosters were counted, along with 38, 120mm mortar rounds, 10 SKS rifles and 66,000 AK-47 rifle rounds. Documents uncovered in a nearby bunker stated that the area was a basecamp for the NVA 86th Rear Service Group.

For the two-day search on May 18th and 19th, Alpha and Echo Companies captured one of the largest caches of the entire Cambodian Incursion. An astonishing three-hundred, seventy five tons of rice and more than 187,000 AK-47 rifle rounds and other munitions were calculated. (It is interesting to note that in the official after action reports and in subsequent works in later years on the Cambodian Incursion, A/5-12 is never specifically mentioned or given credit for the find, only that elements of the 1st Cavalry Division were responsible).

Combat Engineers from the 31st Engineer Battalion were called into the cache site to backhaul the massive find and its contents back to FSB Brown. Bulldozers and Rome Plows had to once again carve another road out of the thick jungle to reach the site.

According to Michael V. Meadows of the 31st Engineers in an interview with <u>Vietnam</u> magazine, "One morning, we were sent into a cache site near FSB Brown. The new road snaked through the jungle, curving and winding and then doubling back on itself as it followed ridgelines and other terrain features. As we inched toward the cache, three NVA soldiers calmly walked out into the roadway. One man squatted down with his rocket-propelled grenade launcher tube across his shoulder. He coolly fired into the side of an APC, blowing several GI's from the top of the vehicle. Return fire roared from the column, quickly dropping the enemy soldiers where they stood. Then, the whole world erupted with the unmistakable 'klack, klack, klack' of AK-47 fire. We quickly realized that a clever ambush had been pulled on us. Because of the meandering route of the road, our own elements were shooting at each other as we tried to establish fire superiority on the ambushers. We kept up a high rate of fire for several minutes until the green tracer fire subsided. With just a moment's hesitation, the column began moving forward again, while dustoff helicopters landed somewhere to pick up the wounded. Later at the cache site, we hauled out 270 tons of rice."

For the next week and a half, Alpha Company continued to guard the cache site and run reconnaissance in force patrols out of the area while the captured goods were moved out via armored convoy or by Chinook.

"During this time," explains Rolf Hernandez, "everybody was eating nothing but rice, rice pudding, rice cakes, rice and C-rations, etc. We grew to hate the little white stuff."

Back at FSB Brown, sack after sack, dump-truck after dump-truck and trailer load after trailer load of rice was being unloaded and slung underneath Chinook helicopters, bound for locations unknown.

Shortly after Alpha found the cache, Reid Mendenhall and his sniper team came back into Brown for a short break before going out on another mission.

As he was walking from the landing zone back to his hootch at the other end of the now bustling firebase, he watched incredulously, "As a low-boy, belonging to an engineer unit, raced past with a woven basket in the middle that looked as though it could have easily fit 30 or more people. It was filled to the brim with rice. The crushed and buckled sides of the bulging container were spilling out thousands of morsels as the low-boy drove on."

"I was the Battalion XO's driver at the time we went into Cambodia," explains Bert Ovitt of HHC/5-12, who was at that time on his third of four combat tours in Vietnam.

"I went out to FSB Brown from Libby some days after the attack there. I arrived late in the afternoon and spent most of the dwindling daylight digging a sleeping trench with an entrenching tool in the dry lakebed. Right at dusk, Major Simmons told me to get a radio and join him on an APC belonging to the 11th ACR while we went out to assist in the recovery of a huge rice cache that Alpha Company found. We left the perimeter with two empty low-boys and an armor platoon. After all of the rice was loaded, I mentioned to the Major that I thought that I should ride back on the trailer which was full of rice bags and baskets, as the trip out there consisted of going down narrow trails where the vegetation came right up to the sides of the APC's. It was dark and I had a very uncomfortable feeling about riding in that particular steel coffin. The Major agreed and I moved some bags out of the center of the pile and built a pretty nice fighting position to ride in. It beat the heck out of being bounced around on the APC. When we arrived back at Brown, they had hot soup ready for us. It was a good thing because it started raining heavily and pretty soon, I was flooded out of my sleeping trench. Fortunately, there was no enemy contact that night."

While Alpha Company was still guarding the cache site, an utterly bazaar incident occurred that is related by Dave Cook of Alpha.

"After spending two or three days there, we were still helping to move the rice out. Throughout the day, helicopters were busy flying back and forth over us and tanks, APC's, bulldozers and low-boys were working like crazy on the ground below. All of the sudden, we started taking ground fire. However, the rounds were coming from one of our own tanks instead of from an enemy AK-47. Incredibly, an unseen NVA soldier had crept into our perimeter and crawled unseen onto one of the tanks that was guarding the cache. The NVA soldier began firing the .50 caliber machine on the top of the turret at everybody around. He was quickly killed before he did any real damage."

Kevin Scanlon had been a rifleman in Alpha Company for several months prior to Cambodia. Scanlon relates an interesting story about the 22nd of May.

"While going to Vietnam, I flew from Oakland, CA, with a guy named Frank Leonowicz from New Jersey. I had been in stateside training with him and several others on the flight at Ft. Jackson, South Carolina. When we arrived in Bien Hoa, we witnessed a huge race riot at the airbase's enlisted men's club our first evening there. Soon after, we were all sent our separate ways with all of the guys I arrived in-country with going to different units. I was sent to A/5-12. Months later, we found ourselves going to Cambodia after spending a short time back in the rear at BMB. After finding the huge rice cache on May 19th, the 2nd platoon was scheduled to go on a recon in force patrol around the perimeter of our find. The date was May 22nd. I remember that I had just received my Red Cross card in the mail that day and one of the guys gave me a roll of 35mm film for my empty Ricoh 35 which I had been humping for weeks without using. I took a lot of pictures that morning of Johnny "Wolfman" Winchester, Raymond "Scarface" Nesbitt, SGT Tony Webber, Jimmy "Redlegs" Setter and our Kit Carson Scout. We left to go out on patrol before I could finish my first roll of film. As we were walking out on patrol, I noticed that the German Shepherd's ears were busy going back and forth. The dog's handler was also straining to hear while looking ahead. Then, all hell broke loose. Moments later, I saw what was left of the dog as it had caught a round in the head. I fired as many rounds as I could at the enemy but we were all pinned down. Raymond Nesbitt was lying to my left and near him was Tanner with his M60 machine gun. Suddenly, as I was firing into the area before me, I couldn't believe my eyes. My M16 had no hand guard and my left index finger was missing. It hadn't gone too far away, only a few inches and it was still connected by what looked like a rubber band. I can't remember who yelled for the medic, but soon, Doc Lonnie Payne was next to me wrapping up my finger with the rest of my hand. Soon after that, I heard someone yell to pull back. We did. As I got up, I noticed that two other GI's were running in the wrong direction. I was somehow very aware of where our whole company was located around that cache of rice. I ran after those two guys and convinced them to turn and run my way. We were the last three guys back into the perimeter. That was my last day in the field.

Scanlon continues.

"I was sent back to the rear with "Hillbilly" (Sp4 Thomas J. Wilder) and "Bravo," (Sp4 Estanislado Bravo) a quiet Mexican guy from Texas to a MASH hospital run by the 1st Cavalry. After some minor surgery, I was flown to the hospital at Long Binh where I was operated on again to reattach my finger. When I awoke from surgery, I looked at the bed next to me and thought to myself that I must be dreaming. Lying next to me was Frank Leonowicz from the 1st Cav. He had busted his leg from jumping out of a Huey. Vietnam was truly the Twilight Zone. Two days later, John "Lucky" Sidange, who grew up in the house across from mine in Hammond, Indiana, was at my bedside. He was stationed at Long Binh and came to visit. A few days later, I was on my way to Japan where I would see Johnny "Wolfman" Winchester come in as I was going home. I spent six months getting operated on every month to save that damn finger. I know how lucky that I was on May 22nd and I will never forget the brave men that I served with; Smitty, Tanner, Joe (for teaching me Spanish), Applejack, Domke, Purcell, Hillbilly and Anderson from Chicago. Where are the now?"

★ ★ ★ ★ ★ ★

On May 23rd, Headquarters Company of 5-12, the TOC, and the battalion's various support units and attached engineer assets, packed what gear they could gather and moved north 10 kilometers further into Cambodia to FSB Myron, which was vacated by the 1st Cavalry's 2nd Battalion, 12th Cavalry. FSB Brown and the memories made there were left to be covered back up by the jungle.

FSB Myron, somewhat larger and more circular than Brown, "Looked like a scab in the jungle." Like Brown, it was situated in the midst of an NVA logistical and supply complex. According to Reid Mendenhall, "Firebase Myron, unlike most bases in Vietnam, had not been sprayed with defoliant and the jungle came nearly up to the perimeter berm. It sat right in the middle of the virgin jungle. There was also a mysterious Montagnard village within 500 meters of the base."

Allen Thomas adds, "When we finally moved to FSB Myron, I thought that since it was on a small hill, there were good fields of fire around, except for the Montagnard village to the west. I don't understand, however, why they would have built a firebase so close to a village. We were ordered that in case of a ground attack, not to fire in the direction of the village. I think that the 1st Cavalry had caused some civilian casualties when they first moved in and we were still in the process of trying to clean up our image to those people."

Situated in the far-flung, triple canopy jungles of southeastern Cambodia, the remote and isolated nether-regions of Central and South Vietnam and the rocky, mountaintop peaks of Laos lived the Montagnards, who were a race of people that were reclusive, shunned, mystical and pre-historic by modern standards.

The Montagnards or, "Mountain People," had for decades, been living and surviving in the utterly remote and near inaccessible regions of Southeast Asia.

"As an indigenous people, the Montagnard were completely different in culture and language from the mainstream Vietnamese. Physically, the Montagnards were darker skinned than the Vietnamese and they do not have the epicanthic folds around the eyes. They are about the same size as the mainstream Vietnamese."

Before the American involvement in Southeast Asia, there were more than one million Montagnards in the region. Throughout their history, the people have always had poor relationships with the Vietnamese, similar to those between the American Indian and the United States government.

Based purely on agriculture, Montagnard culture and economics relied solely on rice farming and slash and burn agriculture.

When Reid Mendenhall of Echo Company first saw them while out on a mission, he couldn't believe that people still lived like that in the 20th Century. "In this village, there were 25 or more natives living there. They were nearly all naked and they lived in five or six different hootches that were elevated off the ground. All of the males carried wooden crossbows that shot arrows or darts that were a little longer than a pencil. The people there looked so bad, that I would assume that they had every disease in the world. Many had open wounds and sores, signs of elephantitis and a few had broken bones from earlier injuries that had healed back incorrectly, looking horrendously deformed. Everything in the village was made of bamboo. Most things were split and tied with vines. They even had different sizes of vines for each job. As we were walking through the village, one of the guys lit up a cigarette with his Zippo lighter. The villagers actually jumped back in astonishment thinking that fire had just come from his hands."

Ihor Dopiwka of Echo Company, like Reid Mendenhall, accompanied the others into the Montagnard villages on several occasions. "All of the women in the village were topless, so there was never a shortage of volunteers to go. One of the other snipers that went with me ended up with one of their amazing crossbows after trading a lighter for it. I can't remember his name, but he was from California and had seen quite a bit of action, having earned two Purple Hearts. He then gave me the crossbow when he went home and then I passed it on down to a good friend of mine when I left also. I remember that the workmanship of the bow and arrows was truly unbelievable. The draw string was made out of strands of vine."

Rolf Hernandez of Alpha Company remembers passing through another Montagnard village and seeing what looked like a dog being slowly roasted over an open fire.

For most of the battalion's time in Cambodia, the medical and civil affairs sections conducted numerous MEDCAP missions for the Montagnards.

"It was the most enjoyable MEDCAP we have ever done. It was good for our morale. Many children were among the patients treated for cuts, dog bites and skin disease. The people were very deserving about seeing the medics and the doctor. They stood in line and took turns. Friendly and unselfish, they showed us a lot of respect and helped us make our mission an enjoyable one," remembers PFC Daniel Smith.

CPT David Kuter, the battalion surgeon adds, "There is quite a large variety of disease which is frustrating because we could only do so much. Everything possible was done to help those people."

CPT David Ashworth, the battalion's S-4 officer, interacted with the various villagers on a regular basis. "Because of all the rice we were finding, we began giving significant amounts of it to the Montagnards as a gesture of goodwill and cooperation. This action, however, had an adverse affect and it was one that we did not realize at the time. Because Montagnard society and culture is based on rice, we nearly started a civil war among some of the villagers when we started distributing the captured rice from the NVA caches. This was more rice than they had probably ever seen in their lifetimes and it immediately started a power struggle. We quickly took back most of what we gave them and sent the rest to Vietnam."

On June 3rd, two weeks after the move from FSB Brown to Myron, Bravo Company, now partially rested and refitted after the May 15th and 16th ordeal, stumbled upon another large find, 20 miles north of Song Be.

(Since the firefight, CPT George Lodoen, who had served as the CO of Echo Company, took control of Bravo after CPT Gordon Lee was reassigned to the rear as the battalion supply officer).

"We were on a RIF (recon in force) toward a suspected enemy cache area in a very hilly region. Our point element had just come down a small hill in an area heavily covered with bamboo when the pointman spotted something ahead of him that looked like a base camp," states Sp4 Calvin Seaman.

After a thunderous barrage or artillery and air-strikes, twenty-seven truck differentials were subsequently uncovered, along with 13 springs,

13 wheels, 35 axles and various other mechanical parts. Later in the day, another part of the motor pool complex was found, which yielded nine jeeps, 12 bicycles and various other truck parts, tools and supplies.

"Unlike the Ho Chi Minh Trail, the Cambodian caches were filled with material transported for the most part, by truck. Truck repair centers and spare parts were a large part of the logistics complex." Well over 300 vehicles were found and destroyed during the Incursion alone. (There was even a 1968 model Porsche and a Mercedes-Benz sedan uncovered in a cache by the 1st Cavalry near "The City" the first week in May).

Because it would have been a logistical nightmare to carry out the pieces of the find, all of the jeeps and most of the tools and parts were blown in place.

"When the demolitions guys blew all of that stuff, we watched, laughing hysterically, as truck parts flew several hundred meters in the air," says Bob Kenna. Needless to say, those parts would most likely never be used again, regardless of the incredible ingenuity and ability of the Vietnamese to salvage and reuse most anything.

That same day, Delta Company, operating 25 miles north of Song Be and close to Bravo Company, also found a sizeable cache of NVA equipment. More than 2,000 bicycles and bicycle parts to build more than 500 more were found in a heavily vegetated and camouflaged depot. The bicycles had been used to carry supplies down the Ho Chi Minh Trail. Heavily modified, the bikes could haul up to more than 300 lbs. or more.

(Two days prior, Delta Company, still the only company in the battalion without any serious enemy contact up to this point, had found 10 sacks of rice laying out in the open several klicks south of this find).

According to the battalion's daily staff journals for Cambodia, the supply and logistical depot found by Delta Company belonged to the NVA's, 86th Rear Service Group, which had been operating from these supply points in Base Area 351 for months.

Delta Company had finally crossed over the border and joined the rest of the battalion on May 25th after spending nearly the whole month on May in blocking positions and conducting patrols around Song Be and Bu Dop.

Bob Pempsell of Delta writes in a letter home on the day Delta crossed into Cambodia, "Well, I finally made it. I'm inside Cambodia and it seems this is where the gooks are. There is action all around us, but the company hasn't seen any yet, although we have found a lot of well-used trails. The Company that I am with is trying to locate enemy caches, but we haven't found anything yet. The hills here are treacherous, man they sure exhaust us from climbing them and they are straight up. It reminds me of the Alleghany Mountains when I used to go hunting there. It sure isn't like you read in the books. My reflexes are getting better all the time and now when I hear a shot go off, I'm down before you know it. It grows on you after a while. I just thought of something. Pretty soon, I will be 21. However, right now, I hear a Cobra Helicopter giving hell to someone with miniguns and rockets going off. Really, the war seems far off unless the shooting starts. Then, it is for real."

The second change of command for the 5-12th Infantry while in Cambodia, occurred at a brief ceremony on June 8th at FSB Myron. LTC David A. Beckner had completed his tenure as commanding officer on May 20th. He was replaced by interim commander, LTC John W. Crancer, who had previously commanded the 3-7th Infantry of the Brigade and was then acting as the Brigade S-3. (When Crancer completed his tenure and returned to Vietnam, he went straight to the hospital and spent a week recovering from malaria at FSB Blackhorse).

The new commander of the Warrior battalion was LTC Wood R. DeLeuil. Having considerable experience, DeLeuil had served a previous tour as an ARVN advisor from 1965-1966 and also as a battalion operations officer with the 101st Airborne in 1969. In the months prior to joining the 199th and after recuperating from wounds received when his helicopter was shot down, DeLeuil was the acting G-3 for XXIV Corps in Da Nang.

Two days after the change of command ceremony, Bravo Company, working in the same area that it had on June 3rd, again hit the jackpot with another treasure trove of NVA goodies. This one, however, took the prize.

Bob Kenna of Bravo Company relates, "On June 10th, we were again searching for cache sites. Operating in rather hilly terrain and going down a hill, we walked into an area that had been hit with defoliant, so most of the surrounding vegetation was brown and dead. We never would have seen the entrances to the storage depots had it not been for that. We counted, I believe, eight tunnels carefully dug into a hillside that were absolutely crammed to the hilt with supplies."

Because the NVA went to great extremes to hide their caches and storage areas, "US forces were not prepared for the difficulty in locating them. Most cache sites were artfully hidden in underground bunkers, deep in triple jungle. There was no pattern to the location of depots and service activities. Some storage sites were adjacent to high speed roads while others were situated in isolated jungle areas, accessible only by foot or bicycle paths."

This site was the largest ammunition and weapons find to date for the battalion. The total of captured enemy equipment was huge: 252,000 AK-47 rounds, 1138 60mm mortar rounds, 45 107mm rockets, 2325 rifle grenades, 6660 radio batteries, 66 75mm recoilless rifle rounds, 80 RPG's, 160 anti-tank grenades, 81 SKS rifles, 4 AK-47's, 11 Chinese radios, 36 boxes of explosive powder, 25 bicycles and one Singer sewing machine was taken.

The next day, other supplies were found, including a cache found by Delta Company, who was then working alongside of Bravo. Delta discovered 61 120mm rockets, 676 grenades and four brand-new K-62 Soviet radios.

According to Bob Pempsell, who as an artillery sergeant was very familiar with radios and communications, "When we found the K-62 radios, the officers were beaming with smiles because they knew it was a super-good find. Those pieces were the latest models that the NVA had."

Because these radios were captured, it indicated that the NVA were hard pressed and did not have the proper time carry out these pieces of high-priority. If that had not been the case, the K-62's would have been the first items to be taken out of the complex.

When taking into account the enormous finds being made by not only the 5-12th Infantry, but the other units involved in the Incursion, the sheer amount of enemy supplies and equipment being found and destroyed was staggering. To say that the North Vietnamese were hurt by the cross-border offensive would be a huge understatement.

More than 11,000 Communist soldiers were killed, wounded or captured in Cambodia from May to June 30th, 1970. Approximately 9000 individual weapons were captured, along with 1200 crew served weapons and 5300 tons of rice.

Although the North Vietnamese and Viet Cong still had adequate supplies of arms, ammunition, equipment and rice in Cambodia and Vietnam, there is no doubt that after the American forces re-crossed the border back into Vietnam on June 30th, the Communist soldiers in the south felt the pinch and sting of the May and June Offensive for the remaining months of 1970 and even into 1971.

(Tom Winfield) FSB Libby, December 1969. Located in northern Long Khanh Province, Firebase Libby served as the battalion headquarters and TOC for the 5-12th Infantry from late 1969 until the deployment into Cambodia.

(David Ashworth) FSB Gladys, 1970. Located astride the Song Dong Nai River in Long Khanh Province, Firebase Gladys was another battalion firebase split between the four infantry battalions of the 5-12th Infantry and Delta Battery of the 2nd Battalion, 40th Artillery. FSB Gladys, as with FSB Libby, was turned over to elements of the 18th ARVN Division when the 5-12th Infantry deployed into Cambodia.

(40th PIO, 199th LIB) Colonel Robert W. Selton, commanding officer, 199th Light Infantry Brigade, April, 1970 to July 1970.

(40th PIO, 199th LIB) A young platoon leader and his platoon sergeant show the strain of patrolling in the triple canopy jungle somewhere in the wilds of Long Khanh Province. Even though a unit could go for days or weeks without an enemy contact, the pressures and hardships of life in the field were always high.

(40th PIO, 199th LIB) "Smoke 'em if you've got 'em." The CP group of Bravo Company takes a break while on patrol near FSB Gladys, 1969.

(40th PIO, 199th LIB) A Redcatcher medic, commonly called "Docs" by the rank and file, cares for a bamboo cut on the head of a fellow soldier. If left untreated, cuts and abrasions made by bamboo and elephant grass, complicated by the filth and nastiness of the infantryman living in the

jungle, could become infected, thus putting a man out of action quicker than a bullet.

(40th PIO, 199th LIB) Delta Company infantrymen somewhere on patrol outside of FSB Brown, Cambodia, June 1970. Oftentimes, the men measured the distance walked by the hills and mountains covered.

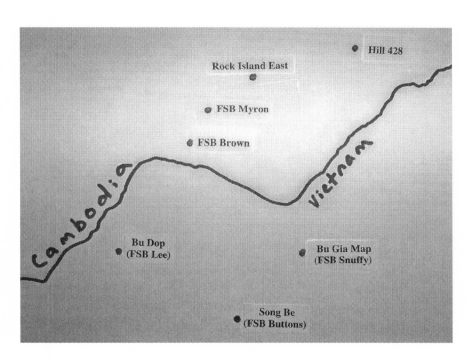

(David Ashworth) 5th Battalion, 12th Infantry Area of Operations, Cambodia, May 12th - June 25th, 1970.

(Reid Mendenhall) The bustling airstrip at Song Be and FSB Buttons, May, 1970. This was the staging area for the 5-12th Infantry going into Cambodia.

(David Ashworth) Fire Support Base Brown, May 25th, 1970. When Bravo and Charlie Companies first landed there on the evening of May 12th, nothing was there save for one strand of flimsy concertina wire and a four foot dirt berm.

(Reid Mendenhall) View taken of FSB Brown's perimeter less than two hours after the firefight ended on May 13th, 1970.

(Reid Mendenhall) Positions occupied by the 3rd Platoon of Charlie Company, 5-12th Infantry during the battle for FSB Brown. The berm and single strand of concertina wire is clearly seen in the foreground.

(Reid Mendenhall) Two dead NVA soldiers from the 174th NVA Regiment the morning after the battle at FSB Brown. These two were shot while running up the berm on Charlie Company' side. It was noticed the morning after the battle that all of the NVA died facing the berm.

(Reid Mendenhall) Beyond the perimeter wire. NVA bodies litter the ground in front of the boundary between Bravo and Charlie Company.

(Reid Mendenhall) Communist weapons and munitions collected from the field on the morning of May 13th. Note the RPG rockets, Chicom grenades and 60mm mortar rounds.

(Reid Mendenhall) Another view of the captured NVA weapons.

(Reid Mendenhall) The only two NVA prisoners to be taken after the battle. These two young NVA soldiers were found in the underbrush within a few meters of FSB Brown's perimeter. They are about to be interrogated by the ARVN soldiers visible in the foreground.

(Reid Mendenhall) The cavalry arrives. ACAV's and Sheridan tanks from the 11th Armored Cavalry and the 31st Combat Engineer Battalion arrive at FSB Brown on the morning of May 13th. They were a welcome sight to say the least.

(Terry Braun) One of the Sheridan Tanks that plowed through the jungle to rescue the remnants of Bravo Company on May 16th. This picture was taken after their rescue and immediately before the ambush on the way out. The thickness of the jungle undergrowth is clearly seen.

(40th PIO, 199th LIB) A Rome Plow in action from the 61st Land Clearing Company. One of these metal beasts was in the lead to rescue Bravo Company on the morning of May 16th.

(Reid Mendenhall) A Sheridan light tank from the 11th Armored Cavalry at FSB Brown. When fired, the vehicles huge 152mm main gun was devastating, especially when using anti-personnel "flechette" rounds.

(Terry Braun) SGT Terry Braun of B/5-12 poses beside a Chinese .51 caliber machine gun and Soviet-made RPD machine guns. This .51 caliber was most likely the one used in the attack on FSB Brown.

(40th PIO, 199th LIB) A small portion of the huge, 375 ton rice cache found by A/5-12 on May 18th and 19th. These sacks have been dumped at FSB Brown where they were either destroyed or distributed throughout South Vietnam.

(US Army) Sacks of rice sitting on bamboo pallets as they were found by A/5-12.

(Malcolm Smith) Two members of B/5-12 examine Communist 60mm mortar rounds after uncovering them at the huge NVA depot found on June 10th.

(Malcolm Smith) Bicycles, used to transport supplies from North Vietnam, are examined at FSB Myron after being uncovered on June 10th.

(Bob Kenna) Bob Kenna (left) inspects a huge pile of NVA 7.62x39 rifle ammunition after finding them buried in caves on a mountainside on June 10th. Kenna served as a riflemen in B/5-12 from 1970-1971.

(David Ashworth) Captain David Ashworth, the battalion's S-3 and S-4 Officer, 5th Battalion, 12th Infantry, 1969-1970. Making sure that an entire infantry battalion operating in enemy territory had everything it needed was a daunting task. CPT Ashworth, by his professionalism and influence, did his job well.

(Terry Braun) SGT Terry Braun, Bravo Company, 5th Battalion, 12th Infantry, 199th LIB, 1969-1970 at FSB Brown. Braun carried this M-203 while in Cambodia, which was an M16 rifle with an attached 40mm grenade launcher. Accurate and reliable, the weapon was introduced in late 1968 and is still in use today.

(Reid Mendenhall) Snipers from Echo Company, 5th Battalion, 12th Infantry at Song Be awaiting transportation into Cambodia.

(John Hart) Sp4 John Hart (center) of Delta Company, 5th Battalion, 12th Infantry takes a break while on patrol. Note the Pepsi cans being used as field expedient stoves to warm up their cans of C-Rations.

(Rolf Hernandez) Sp4 Rolf Hernandez of Alpha Company, 5th Battalion, 12th Infantry. Hernandez discovered a hidden cache of SKS rifles on May 18th, which was in the same vicinity as the 375 ton rice cache.

(Jim Horine) SGT Jim Horine, Infantryman and RTO, Bravo Company, 5th Battalion, 12th Infantry, 1969-1970.

(Jim Horine) Jim Horine on the afternoon of May 12th, 1970 less than an hour before going into Cambodia.

(Reid Mendenhall) SGT Reid Mendenhall, Charlie and Echo Companies, 5th Battalion, 12th Infantry. After serving as a grunt in Charlie 5-12, Mendenhall joined the sniper teams of Echo Company and became a team leader. Note the highly-modified M21 sniper rifle with the Redfield ART-1 3x9x40 scope.

(Bob Pempsell) SGT Bob Pempsell holding the SKS he claimed and took home from Cambodia. SGT Pempsell, Airborne qualified, was an artillery reconnaissance sergeant from Delta Battery, 2-40th Artillery attached to D/5-12 in Cambodia and after.

(Dick Rose) 1LT Dick Rose (center), Alpha Company, 5th Battalion, 12th Infantry at the huge rice cache found on May 18th and 19th. Rose had already spent six months in the field as a platoon leader before the battalion went to Cambodia. Coming back from R&R the first week in May, Rose was given command of the 1st Platoon in Cambodia after the platoon leader was wounded in action.

(Malcolm Smith) SSG Malcolm Smith, Bravo Company, 5th Battalion, 12th Infantry, 1969-1970. SSG Smith, highly-respected by the men that knew him for his knowledge and experience, served three combat tours in Vietnam, the first being with the 4th Infantry Division and his last with the Vietnamese Rangers. His second tour was with the 199th LIB.

(Malcolm Smith) SSG Malcolm Smith (left) and another member of Bravo Company shortly before leaving for Cambodia.

(John Wensdofer) Sp4 John Wensdofer, Charlie Company, 5th Battalion, 12th Infantry. Arriving as a replacement in the field for Charlie Company in February of 1970, Wensdofer was wounded in action less than twenty-four hours later. On May 13th, he and his red-hot M60 placed a withering fire for over three hours on the fanatical NVA attacking FSB Brown.

(Reid Mendenhall) The Snipers of E/5-12 shortly before moving into Cambodia. Ihor Dopiwka (center) listened to the last words spoken by WO1 Robert Gorske after being hit near Hill 428 on May 21st, 1970.

(Reid Mendenhall) Snipers from E/5-12 pose with their M21 rifles at FSB Libby. The soldier on the right has a Starlight Scope mounted on his rifle.

(Reid Mendenhall) A moment of reflection. A member of E/5-12 is lost in his thoughts on the way to Cambodia, May 13th, 1970. None of the men from the battalion had any idea what they were getting into.

*(Bob Pempsell) CPT Mack Gwinn of Delta Company (left) and SGT
Bob Pempsell (right) while in the field near Tanh Linh, 1970.*

(Bob Kenna) One of the Kit Carson scouts attached to B/5-12. This one was killed in action on May 18th.

(David Ashworth) Montagnard tribesman from one of the small villages that dotted the jungle in between FSB's Brown and Myron. Seen as outcasts by the Vietnamese and other Asian ethnic groups, the Montagnards lived in some of the most inaccessible regions of Southeast Asia.

(US Army) A dog handler and his German Shepherd searching through one of the many NVA base camps near FSB Brown. The men and dogs from 76th Combat Tracker Team provided invaluable assistance to the 5-12th Infantry while in Cambodia. One of the team members was killed in action on June 16th while on point for Delta Company.

(Bob Pempsell) An M102 Towed, 105mm howitzer from Delta Battery, 2nd Battalion, 40th Artillery at FSB Gladys.

(Reid Mendenhall) A cannon-cocker from D/2-40th Artillery shows the strain of constant fire missions while at FSB Brown, May 1970. The gallant and professional artillerymen of 2-40th Artillery averaged firing over 650 rounds a day while in Cambodia.

(Bob Pempsell) A 1st Cavalry Division AH-1G Huey Cobra from the "Blue Max" Squadron. Providing close air support to the 5-12th Infantry, the Cobra was armed with a 40mm grenade launcher, an XM134 mini-gun, wing-mounted 7.62 machine guns and 2.75 inch rockets.

(Bob Pempsell) Close up shot showing the wing-mounted armament system of a AH-1G Cobra. Shown here is the XM134 mini-gun and the 2.75 inch rocket pods.

(Tom Winfield) An OH-6A Cayuse, "Loach" light observation helicopter from Fireball Aviation, similar to the one that WO1 Robert Gorske and SGT John Rich went down in on May 21st, 1970.

*(Bob Pempsell) A CH-47A Chinook helicopter from the 1st Cavalry
Division. The Chinook was a genuine workhorse used to transport
weapons, artillery, vehicles, ammunition, food and up to 33 troops. Most
men from the 5-12th Infantry rode into and out of Cambodia in one of
these machines.*

(Bob Pempsell) The Chinese SKS rifle and the US Colt manufactured M16A1. Because the SKS was a semi-automatic weapon, it was allowed to be taken home as a trophy. Most of the men from the 5-12th Infantry that served in Cambodia were able to send one home.

(David Ashworth) FSB Myron as seen from the air, June 1970. The firebase was often referred to as "A scab in the jungle."

(David Ashworth) One of the many Montagnard village that dotted the landscape between FSB Brown and Myron.

(David Ashworth) View of one of the Engineer roads constructed to backhaul the contents of the huge NVA cache found by Bravo Company on June 10th. Note that the jungle vegetation stretches right up to the edge of the road created by the Rome Plows.

(Jim Horine) The 3rd Platoon of Bravo Company, back from Cambodia at Song Be, June 25th, 1970. Tired, haggard, filthy but alive, the survivors of the battalion went back to Camp Frenzell-Jones for a much-need week stand-down. Then, it was back to the war for most.

Ch. 4

HILL 428

(May 21st, 1970)

The world is a bad place,
A bad place, a terrible place to live.
Oh, but I don't want to die.
All my sorrow,
sad tomorrow.
Take me back to my own home.
All my crying,
Feel I'm dying, dying.
Take me back to my own home.

"Reflections of My Life"
The Marmalade, 1970

"May 21st, 1970. Just another day to mark off on my short-timer's calendar. Ninety-two more days and a wakeup to go and then I'm home free. On that Freedom Bird, out of this place, out of this situation and out of the Army. That's freakin' great."

As I roll up my poncho and poncho-liner and cram it into my already bulging rucksack, I think to myself about what I have been through the last seven months of my life.

"Vietnam was bad," I think. "Cambodia has been an utter bitch. Man, what a roller-coaster ride this experience has been. How will I ever be able to forget about this place, the people I have met, the people I have killed, the sights, sounds, smells, memories, nightmares, fear and excitement. What will I tell my wife when I get home?"

I open a can of dented and leaking fruit-cocktail, gulping the dull tasting sustenance down before Charlie Company again moves out on another patrol. I am still lost in my own thoughts.

My hands and arms are covered in sores and bamboo cuts. My feet are nothing but blisters and I haven't shaved in over a week. I believe that I have lost over 25 pounds and I haven't had a full-night's sleep since my tour of duty began. I smoke too much and my nerves are wearing thin and I am always on edge.

Since the horrific firefight at LZ Brown, we have been in contact with the NVA every day since. Thankfully, we have not had any serious fighting, save for a lot of sniper fire, hit and run ambushes and several mortar attacks. That all changed yesterday afternoon. The NVA hit us again from less than 50 meters out with heavy automatic weapons and RPG fire, when we bumped into approximately 10 of them while we were searing for more cache sites.

That guy next to me, SGT John Menefee, was wounded in the arm by an AK round. It severed the artery and he nearly bled to death before the medic stopped the bleeding and the dust-off came in to pick him up. Sp4 Robert R. Nutt, Sp4 James W. Randolph, PFC A.C. Cooper, PFC Ronald F. Martin and Sp4 Michael J. Sutyak were also hit. We did however, kill five of them. Just a few minutes later, the 2nd Platoon got into contact on their side when they saw over 20 NVA moving down a trail towards them. The guys in the 2nd Platoon then withdrew back to the rest of the company because they were outnumbered.

"The NVA are all over this place. I wonder if today, May 21st, will be any better than the rest?"

Hill 428, so named because of its elevation in meters above sea level, sat in treacherous, Communist-controlled terrain approximately seven klicks northeast of FSB Brown and 26 miles north of Song Be. The area was literally crawling with NVA who had been unhappily ousted from their basecamp at the huge arms and logistical complex called "Rock Island East," several days earlier by elements of the 1st Cavalry Division.

The area was much too hot and active for one American line company to be working in, but on May 21st, Charlie Company, then commanded by CPT David Thursom, was pushing extra hard to cover more ground and find more enemy supplies.

CPT Thursam, who had been in command of Charlie Company since February, was "well-liked and very much respected" by the men in the company. He however, was under a lot of stress at this time and was being pushed incessantly by LTC DeLeuil, who in turn was being ruthlessly prodded by COL Selton at Brigade, who was in turn being given no quarter by the 2nd Brigade, 1st Cavalry Division, II Field Force, MACV, etc., etc. In short, from platoon leaders on up to major generals, the grind and intensity of the campaign was being felt.

(The momentum and frequency of finds in the Cambodian Operation up to that point in May were seriously picking up and the American high command was adamant about covering as much territory and destroying as much of the enemy's logistical complex in Cambodia as possible before the June 30th deadline).

In the spring of 1970, Charlie Company of the 5-12th Infantry, 199th Light Infantry Brigade was fairly typical of the hundreds of American line infantry companies operating in Vietnam and Cambodia.

Composed of a unique mixture of whites, blacks, Native Americans and Hispanics, the average age of the combat infantryman in Charlie was 20.5 years old. Nearly all had graduated from high school and a substantial number of them had earned college credits or had attended trade schools before being drafted. Several of the "older" men were even four-year college graduates.

Over half of the company had been in-country from four to six months and were seasoned, combat veterans by the time the battalion was deployed to Cambodia. Although they had not experienced the type of large-scale, daily contacts like they encountered across the border, they had fought in their fair share of vicious and deadly firefights in Vietnam. Since January, the company had lost four of its members killed in action, with well over ten times that number wounded.

Like the three other infantry companies in the battalion, Charlie Company had been humping and sweating in the same area of operations around Long Khanh and Binh Thuy Provinces, out of FSB's Libby and Gladys, for most of late 1969 and all of early 1970.

"While there," states Allen Thomas of the battalion, "we had a very fair rotation of companies going into and coming out of the field on missions. It took one company of infantry to adequately guard the firebase. The 5-12th Infantry allowed the four companies to rotate a new company into the firebases every three days, allowing time for the men coming off a nine day mission to clean weapons, zero weapons, take on new equipment, get a clean uniform and have some hot chow and drink all the beer you could handle. Of course, it doesn't always work perfectly. The longest mission that I was ever on lasted for 15 days. By that time, our fatigues were actually rotting off our bodies."

When Charlie Company began humping towards Hill 428 on the morning of May 21st, they had already been out in the thick, hostile jungle for well over a week. The grunts were tired, sore, hot and homesick. Many were physically sick, suffering from fever, dysentery, jaundice, exposure and from the early symptoms of malaria. The men were looking forward to going back to FSB Brown on "Palace Guard" for a couple days of showers, sleep, beer and letter writing while guarding the now bustling firebase perimeter.

By early afternoon, the point squad was paralleling a high-speed trail that was going in the same direction as their route of march. The men were strung out in a long, meandering company file that snaked lazily through the green foliage and bamboo. The order was given to close it up as the point squad and those following immediately behind, were painstakingly hacking their way forward through the wait-a-minute vines and bamboo with machetes, all the while trying to be a silent as they could. From the jagged, slopes of Hill 428, the NVA knew the Americans were coming.

Somewhere off to the west at around 1230 hours, for those able to hear that weren't panting or out of breath, came the deep boom of artillery fire and the chilling rumble of an air-strike. Just a few clicks away, somebody was in contact. The enemy was close.

"Damn it's hot," I utter to nobody in particular. Struggling to get a canteen out of it's pouch, I take a quick gulp of the warm, foul-tasting liquid. "These iodine pills may make this stuff clean to drink, but it doesn't do shit for the taste."

I put the canteen back in its pouch, grip my rifle tighter in my hands and take a few more labored steps through the bamboo when the column suddenly stops. Some of the guys around me start to light up cigarettes while others collapse to the ground under the weight and strain of their rucksacks, weapon and ammunition.

The word begins to filter back that the point squad cannot go forward another step. The jungle is just too thick to bust or chop through.

Garbled voices begin to excitedly crackle from transmissions over the PRC-25 radio that the RTO is carrying behind me. I study his face as he is listening in on the conversation and visibly see that his expression suddenly changes to one of utter astonishment.

"We're going to back up and get on the trail," he loudly whispers to the squad leader who is looking back at him in equal shock.

John Wensdofer was back at FSB Brown, suffering painfully from diarrhea, severe jaundice and a high fever on May 21st.

Regarding the decision either ordered to or made by Charlie Company to walk down a heavily-traveled enemy trail, Wensdofer relates, "I believe that CPT Thursam was ordered to get out of the bush and on the trail by either LTC Beckner, COL Selton or someone else. After all, the brass was flying above their position in C&C ships. CPT Thursam would have never have told the men to do this. We had always, and I mean always, avoided walking on trails before May 21st."

Roger Lowery of the S-4 section also echoes Wensdofer's statement. He says of CPT Thursam, "I would have gone out with him most anywhere. He was well-liked and a very experienced commander. I have a hard time in believing that he would have ordered his company to walk on the trail."

Despite the command, the grunts in Charlie Company reluctantly obeyed and stepped gingerly on the well-used trail, astonished by the hundreds of sandal and shoe tracks made by the NVA. The time was 1410 hours. The young pointman cautiously looked from side to side, flipped the safety of his M16 and slowly began walking forward.

The noise was deafening and the volume of enemy fire was unlike any heard before. "It seemed like a million fireworks going off at once," states John Wensdofer who was listening to the firefight at Brown. "It was terrible."

Charlie Company had entered into the threshold of hell by walking into a near-perfect U-shaped ambush at the base of Hill 428. The Warriors were outnumbered and outgunned from the very start, receiving devastating torrents of fire from AK-47's, SKS carbines, RPD machine guns and RPG-7's from their north and northwest.

SGT Warren L. Scanlon, 21, married and Sp4 Donald G. Busse, 20, were immediately killed in the first few seconds of the ambush. Their bodies, as they were walking point, were lying together several yards ahead of the now pinned-down and bleeding column.

Sixteen other soldiers were seriously wounded within minutes of the NVA springing the ambush. (PFC Harry J. Britton, PFC Paul E. People, 2LT Michael S. Wise, SP4 Michael J. Kulpa, PFC Stephen C. Vicko, Sp4 Jerry D. Ortel, SSG Timothy J. Osborn, Sp4 Woodrow Holleyman, Sp4 Luther W. Moore, Sp4 Alan J. Schmitt, PFC Tony Castillo, PFC Richard C. Moore, SGT Joseph R. Speranza, PFC Dan P. Serrano, Sp4 Garfield Stanley and PFC Bert Winn).

With the projectiles of death and disfigurement swirling about them, the heroic Combat Medics, or "Docs" as they were commonly called, worked feverishly to stop the bleeding, prevent shock and patch up the broken bodies of the wounded men. By day's end, the medics, out of bandages, plasma and most other critical supplies, bound the wounded men with socks and strips of fatigue jackets that had been ripped apart to serve as makeshift bandages and tourniquets.

The murderous fire coming from the slopes of the hill had not slackened or subsided, making it nearly impossible for any of the now pinned-down GI's to move. The North Vietnamese had the distinct advantage of elevation and fire superiority.

Despite the noise, bullets and shrapnel exploding and hissing around him, CPT Thursam formed what was left of the company into a loose perimeter several meters back from the kill zone while the NVA tried to surround and outflank them. The bodies of Scanlon and Busse were left where they had fallen. (They would be retrieved several days later on June 9th).

It was now 1430. Once together inside the perimeter, those members of Charlie Company who were not wounded or still in a state of shock from the enemy fusillade, began pumping out return fire, shooting at anything and everything around them. It was enough to keep the NVA at bay for the time being.

Roger Lowery had joined S-4 supply section of the 5-12th Infantry in June of 1969 when the Brigade was still in the beginning stages of moving from the steamy rice paddies of the Pineapple region south of Saigon to the mountains and jungles of Long Khanh Province. Lowery's primary responsibility was to re-supply line units in the field, anytime and anyplace, by helicopter. He had fifty days to go in country when he went into Cambodia with the unit.

According to Lowery, "When a unit got into contact, my job was to load the available air assets with as much ammunition and supplies the helicopter could handle and get the items out there to the guys that needed it as quickly as possible. When Charlie Company got into contact on May 21st, they were out of ammunition and first-aid supplies within minutes of the first shot being fired. That is how intense that particular firefight was."

"When the North Vietnamese started firing, I remember thinking that it was far heavier than at FSB Brown. I don't think I can describe how intense the firing was for the first several minutes, until we pulled back. I wanted so badly to shrink and be able to fit inside my helmet. Enemy rounds were hissing and popping all around me. I could actually feel them passing by my face as I lay on the ground, trying not to move. Despite all the noise, I could still hear the sickening sound of bullets and shrapnel striking flesh and bone. How was it that so many around me were hit and I was untouched? When we finally pulled back and formed a perimeter, I started loading and firing my weapon on automatic as fast as I could. I shot at anything that was in my front. I have no idea if I hit anything as the NVA were superbly-camouflaged. Within minutes, I was down to my last five magazines."

Inside the embattled perimeter, CPT Thursam was in contact with LTC Beckner and the TOC back at Brown, urgently requesting a medevac and air and artillery strikes. The company was still pinned down, not knowing if the NVA were going to try and overrun them with one massive attack.

By 1600, Dustoff 23 was on station over Charlie Company's position, waiting for a lull in the firing so that it could extract some of the wounded with a jungle penetrator.

Finally, after the fire had slackened somewhat, the slick swooped in and hovered over a small hole in the jungle canopy. It was met by heavy ground fire, forcing the helicopter to break station. Several minutes later, a second attempt was driven off also.

On the third and last attempt, Dustoff 23 was able to life out three fortunate soldiers who were seriously wounded. It was to be the last attempt made to get the injured soldiers out for the next 16 hours.

Back at FSB Brown, the gun crews from Delta Battery, 2-40th Artillery kept round after round after round of HE slamming into the North Vietnamese who were still aggressively trying to close the noose around Charlie Company from the slopes of Hill 428.

The 105mm rounds, crashing through the trees and jungle canopy, burst into thousands of pieces of hot, jagged metal, taking a toll on the enemy. Unable to maneuver closely on the GI's from this ring of steel, the NVA continued to molest the Redcatchers with small arms and RPG fire.

By 1700, CPT Thursam, still in contact with the TOC, requested an immediate re-supply of ammunition, first-aid supplies, bandages and smoke grenades. (Thursam had also nearly run out of smoke and desperately needed more to mark his position for the fast-movers that were inbound carrying napalm).

Back at FSB Brown, WO1 Robert E. Gorske of the 199th's, Fireball Aviation heard Thursom's call for help and immediately cranked up his UH-1 Huey helicopter. In-country since February, Gorske, along with

SGT Roger Lowery, crammed what boxes of ammunition and grenades they could into the aircraft and buzzed off into the fray. (The crew-chief/gunner on Gorske's helicopter refused to go with them. Gorske and Lowery were the only ones at that time to go on the re-supply mission).

In seconds, Gorske and Lowery were hovering over Charlie Company's position, requesting that they pop smoke so they could identify their position and get on with the drop. According to Lowery, "When we were over their general location, purple smoke started drifting upwards through the canopy. As we were looking down on this location, another clump of purple smoke started to also rise to our left in another part of the contact area. The NVA, watching what was going on, popped their own smoke. I was not sure that we were over the right area until a pitiful looking American soldier appeared in the middle of the rotor wash below." With that sight, Lowery ferociously began unloading the desperately needed supplies as fast as he could.

AK-47 rounds were now punching half-dollar sized holes through the aircraft, passing mere inches from Lowery's face. WO1 Gorske banked and yanked the collective to the left and scooted back to Brown for unbelievably yet another re-supply of ammunition and grenades.

After quickly shoving more ammo and supplies into the slick back at Brown, they were back over the raging firefight within minutes.

Despite the contact area still being too hot, Gorske again hovered over the perimeter while Lowery frantically kicked out the supplies. Lowery continues. "On the second drop, we took more incoming rounds. I remember the noise of the battle and the screaming conversation that I had with Gorske. I was yelling at him, telling him to get in position and checking to see if there was an LZ anywhere in the are where we could land and try to get out some of the wounded. He was yelling back at me that we were taking serious ground-fire and being hit. Out of frustration, I pulled out my captured Chicom 9mm and emptied the magazine into the jungle below."

Still airborne, despite more than fifty bullet holes punching through fuselage, the helicopter limped and sputtered back to the LZ at Brown where Gorske hit the ground hard and shut the engine down.

As Gorske exited his machine, COL Carter Clarke, the commanding officer of the 2nd Brigade, 1st Cavalry Division landed in his own helicopter to monitor the situation on Hill 428.

Gorske approached COL Clarke's "Loach" or Light Observation Helicopter (Tail # 67-16454) and recognized WO1 Patrick F. Cawley of the 1-9th Cavalry. The two had gone to flight school together at Ft. Rucker, Alabama, prior to going to Vietnam.

Talking Cawley into going with him for a third re-supply attempt, the two pilots prepared the bird for takeoff while SGT John W. Rich, Charlie Company's supply NCO, left his post at the firebase and boarded the aircraft to fly back with Gorske and Cawley. (Shortly after landing at Brown after the second drop, Lowery was ordered by both CPT David Ashworth and COL Selton not to go back over the contact area for another attempt).

John W. Rich, 21, married, had already spent the past 11 months in the field humping as an infantryman with Charlie Company. On May 21st, Rich had three weeks to go in country before ending his tour of duty and going home.

John Wensdofer, who was friends with Rich, remembers him as a super-guy that was always willing to help out the others in the company. "When he got on the chopper that day, he didn't have to be there, but he did it any way. Our eyes met for the last time. I looked at him and he at me. By his expression and gaze, I believe he knew that he was going to die."

The time was 1825. For the third and final act, Gorske buzzed over the treetops and hovered over the same place in the jungle that he had two times before. The grunts and artillerymen back on Brown stood up on the berm and those with binoculars watched intensely to see if Gorske would cheat death once again. The North Vietnamese, however, were waiting.

The small but rapid-moving reconnaissance helicopter reached the hole in the canopy and again came to a hover. It was there for no longer than five seconds, "When it looked like an invisible hand began to smash and shake it in midair." The Loach was nearly blown apart by the force of hundreds of enemy rounds, stitching the helicopter from one side to the other. The Plexiglas shattered with tremendous force and other parts of the body were ripped away from the fuselage. The grunts on the ground stared in awe as the helicopter dipped and the engine momentarily stalled, then restarted, gained a little bit of altitude and like a smoking, wounded bird spiraling out of control, flew back towards FSB Brown.

Ihor Dopiwka and the rest of his sniper team were hidden in the bush two klicks north from Brown and had been listening to the sounds of the battle since it started. Since Dopiwka carried a PRC-25, he had been listening to the conversation between Gorske and the TOC.

"I was listening in on the events going on while Gorske and his helicopter went to re-supply Charlie Company for the third time. When they were hit by ground fire, you could actually hear the bird being blown away on the handset. It was tragic. I don't know if it was Gorske or not talking, but I remember what was said in the last few seconds. Mayday, Mayday, we've been shot up bad. The pilot and passenger are both dead. I am going to try and make it back to Brown. The guy in the TOC was offering words of encouragement to the pilot saying, "Keep it up, keep it up, you're almost home," to which the pilot of the helicopter replied, I've got Brown in sight. Then, there was nothing, save for the voice of the duty officer back at the TOC at Brown frantically calling for a sitrep."

With 500 meters left to make it back to Brown, the tail boom on the Loach suddenly tore off, causing the craft to immediately nose-dive into the trees. There was a small explosion, followed by thick, black smoke curling upwards through the jungle canopy. (The helicopter went down at coordinates YU080383).

Back on the perimeter at Brown, the men watching the scenario unfold dropped their heads in silent despair. CPT George Lodoen, still the commanding officer of Echo Company at this time, had been watching intently through his binoculars. He shouted at one of his sniper teams to, "Gear up, we're going in to get them!"

Reid Mendenhall was on the sniper team summoned to go on the rescue mission. "When we heard the call for help from CPT Lodoen, we gathered our weapons and gear in great haste. As I recall, two sniper teams, a couple of medics and some infantrymen from other companies answered the request."

John Wensdofer of Charlie Company, who missed the firefight going on because of a high-fever and diarrhea, also joined the rescue team.

While CPT Lodoen was getting his people together, a Cobra gunship and two lift ships (Hubble 44 & 65) carrying a "Blues Team" from the 1st Cav's, 1-9th Cavalry also reported on station over the now smoldering crash site.

"Within minutes, we were ready," says Reid Mendenhall. "We left the firebase and literally started charging through the jungle towards the crash site at a dead run. We really were not paying attention to what we were doing because we were so focused on getting to the helicopter. I don't even think that we had any security out watching our flanks."

Several hundred yards into their cross-country run, the officer in charge of the circling aero-rifle team radioed CPT Lodoen and told him there was no movement below and no sign of survivors.

By 2005 hours, CPT Lodoen and the rest of his detachment had reached the crash site. The sunlight, however, was fading fast into darkness. As the day came to a close on May 21st, 1970, Lodoen and his snipers melted into the bush a few meters away from the crash to set up a small night defensive position while Charlie Company, still surrounded and under fire from the NVA on Hill 428, hunkered down in the middle of enemy territory with their 13 wounded.

For the next ten hours or so, the Cambodian jungle around the hill was violently and suddenly punctuated by brief but deadly spurts of automatic weapons fire, the hollow boom of fragmentation grenades, the shriek of incoming artillery and the roar of enemy RPG rounds. Flares were kept illuminating the perimeter all night with their eerie hiss illuminating the ghastly landscape.

From another hilltop several kilometers to the east, the cavalrymen from Bravo and Echo Companies of the 1st Cavalry's, 5th Battalion, 7th Cavalry had heard the ferocious firefight going on at Hill 428 for most of the day. That night, they watched the flares dance lazily in the sky as tracer rounds bounced and ricocheted into the air. They also saw headlights from the trucks of an NVA convoy leading away from the landmass. It was assumed that the NVA were protecting something big. They were.

Finally, the warm morning came on May 22nd and with it, the ever-present prospect of either living or dying on another day. Sometime during the night, the North Vietnamese withdrew and loosened the chokehold around CPT Thursam and the survivors of Charlie Company.

Pulling back several klicks further and clearing an LZ in the jungle just large enough for one Huey to come in at a time, Charlie Company was extracted later that day. A total of 21 young men were wounded and four were killed in action, counting WO1 Gorske and SGT Rich. There were so many wounded, that a CH-47 flew in and evacuated them to Song Be.

Once again approaching the Loach's crash site, CPT Lodoen and his men began the grim task of identifying the bodies and collecting their remains. Reid Mendenhall explains, "The small helicopter was approximately 30 feet off the ground, hung up in the trees and badly burned. When we finally got it down, we noticed that it had been stripped of any weapons or equipment."

While the snipers and attached infantrymen provided security, the two medics stoically began picking up the remains of the crew, gently placing them in body bags. Two of the bodies, burned beyond recognition, were still strapped in their harnesses.

John Wensdofer emotionally recalls, "We returned to FSB Brown with the bodies that evening. It was May 22nd, my friend John Rich's birthday. He was to be 22 years old."

Roger Lowery adds that, "What occurred in Cambodia on May 21st, 1970 is the single most incident in Vietnam and Cambodia that I think about the most, everyday. That should have been me on that last drop, not Jon Rich. He was doing my job and he died in my place. I owe my life to CPT David Ashworth. Robert Gorske deserved nothing less than the Distinguished Flying Cross or more. He was extremely brave and he went into that area three times to give his brothers-in-arms on the ground a life and death supply of ammunition and supplies. Both Gorske and Rich did not have to do what they did and they paid with their lives. To my knowledge, they have never been awarded or recognized for their sacrifice."

The NVA began firing from the slopes of Hill 428 again during a heavy rain that afternoon. After Charlie Company of the 5-12th Infantry was extracted, Bravo and Echo Companies of the 5-7th Cavalry, which had been humping in the direction of the hill since the day before, took up their slack and picked up where the Redcatchers left off.

As they crept closer to the base of the hill, the "Skytroopers" passed through the area where Charlie Company had fought the day before. Bloody bandages, mountains of shell casings, empty magazines and other articles were strewn across the contact area.

Cautiously, the GI's swept forward, now beginning to climb up the hill. "About halfway up, at a five-foot ledge, we began getting AK-47, machine gun and B-40 rocket fire," said SGT Pat McConwell of E/5-7. "They could have rolled grenades down on us from up there. I'm sure glad they didn't."

A furious firefight ensued and Bravo and Echo Companies pulled back to the base and called in air and artillery strikes for the rest of the day and into the night. "I called in artillery all night," recalls 2LT William Harrington of Echo Company, "and the next morning, the Air Force came in again and pounded the hell out of the top of that hill."

At FSB Brown, while the cavalrymen were battling up the hill, the six guns from the 2nd Battalion, 40th Artillery continued, as they had for Charlie Company the day before, sending round after round crashing into the NVA dug into the top of Hill 428.

As the breech was closed on Gun #6 and the lanyard pulled to fire, the howitzer inexplicably exploded, sending searing pieces of metal and shrapnel all over Brown.

PFC Daniel E. Nelms, 21, the assistant gunner, was killed instantly while four others from the crew were seriously wounded. The rest of the other gun crews stopped firing immediately. Nobody knew what had happened and since they were mortared and sniped at frequently, the first conclusion was that the explosion was due to some type of enemy fire.

Ihor Dopiwka of the Sniper teams witnessed this horrific calamity. "When the 105 blew up, I was right there. We, the sniper teams, were getting briefed for another mission when the gun unexpectedly exploded. I remember the tremendous blast and seeing half of a howitzer barrel flying through the air, all the while on fire. I thought to myself how odd it was that steel could burn like that. Naturally, we ran over to help but their was nothing that we could do."

When the confusion died down somewhat, 1LT Stan Hogue, Delta Battery's executive officer, noticed that the explosion probably came from a defective 105mm shell.

Because they were in the middle of a vital fire-mission and because there were infantrymen in contact depending on them, Hogue, along with one of his NCO's slammed a round into the gun beside the one that had just been blown to pieces and sent the shell on its way.

Reassured, the rest of the gun crews went back to work and finished the mission. (A later investigation did in fact reveal that a defective shell had exploded in the breech of the gun, thus causing the calamity).

Of all the unsung heroes that fought in Cambodia, the artillerymen of Delta Battery deserve to be right at the top of the top of the list. For two long, arduous months, the men on the guns averaged firing several hundred rounds a day, getting less than 2-3 hours of sleep in a 24 hour period. Even the grunts out in the bush on patrol had more time to relax on a daily basis (when not in contact) than the redlegs.

Thousands of soldiers owe their lives and well-being to the dedication and professionalism shown by these cannon-cockers of the 2nd Battalion, 40th Artillery from 1966-1970. After each infantry company returned to either FSB Brown or Myron after days in the field, most of the men would visit the gun sections of Delta Battery and personally thank them for a job well done. (While in Cambodia, Delta Battery fired more than 29,000 rounds in support of combat operations, averaging more than 650 rounds a day).

Finally, on May 23rd, after another long firefight in the pouring rain, Bravo and Echo Companies of the 5-7th Cavalry took the hill after the NVA on top again melted back into the jungle.

It had taken three days of near continuous and heavy fighting and the lives of six American soldiers from the two units to capture. However, it soon became apparent why the North Vietnamese had fought so hard to protect it.

Hill 428, now called Shakey's Hill after a soldier killed on May 23rd from the 5-7th Cavalry, turned out to be another huge cache site.

Bunkers, dug in as far as 12 feet into the ground, contained pallet upon pallet of small arms and ammunition. In all, more than two-hundred tons of supplies, including several Russian-made flame throwers, were captured there. The site was so extensive that a small firebase was constructed there by elements of the 3rd Brigade, 9th Infantry Division solely for the purpose hauling out the captured enemy munitions and supplies found there.

Two weeks later on June 8th, the hill was bombarded by generals from MACV, congressmen from Washington D.C. and reporters from all of the major media outlets on a presidential fact-finding tour, complete with Air Force jets making fake bomb runs, inspections of the enemy bunker and logistical complex, detailed briefings and captured enemy weapons and equipment being shown off.

Seventeen of the highly-prized and sought after SKS rifles were given to all of the eager to receive congressmen as souvenirs. According to David Kuter, the battalion surgeon for the 5-12th Infantry, "They (the politicians) claimed seventeen of the weapons that had already been assigned to individual soldiers. Thus, the men who sweated out here for days at great risk to their life had the one souvenir they earned taken away just so a bunch of thoughtless politicians would have something to hang on the wall and brag about how they were in Cambodia. Events like that really make one grit his teeth."

Forty-eight hours after their near fatal experience on Hill 428, Charlie Company was sent back across the border to Song Be and Mt. Thomas for a three-day stand-down.

John Wensdofer writes in a letter home to his wife Dianna on May 24th. "Well, I'm back in Vietnam and really glad to be here. Boy, I never thought I would ever say that. My company is now down to 57 men and we are really tired and beat. Right now, we are on the top of a huge mountain, built up just like a firebase. We will stay here for a few days and then go back in until June 30th. There are so many NVA in Cambodia that no matter were you go, you are sure to make contact. Last night, before we left to come here at Firebase Brown, we were mortared again. Luckily, nobody was hurt. That's enough about the war."

Ch. 5

DELTA COMPANY'S DILEMMA

(June 16th - 23rd, 1970)

"I hope that I don't ever have anymore days like that again."

Bob Pempsell, D/5-12

"Hump, hump and hump. That's all we ever seemed to do. Go out on patrol and walk through the jungle," relates John Hart of Delta Company.

While the rest of the battalion entered Cambodia on May 12th and 13th, Delta Company had remained behind in Vietnam, patrolling and conducting reconnaissance in force missions around FSB's Snuffy and Buttons, while also providing security around the bustling airstrip at Song Be.

Contact with the Communists had been relatively light while there, as compared to the other companies in the battalion that were operating across the border. That would change drastically for Delta toward the middle of June, however.

CPT Brice Barnes had commanded the company since February, 1970. A multi-tour soldier, Barnes was an extremely experienced commander, having earned the Distinguished Service Cross with the 2-47th Infantry of the 9th Infantry Division during the Tet Offensive of 1968. (Ironically, the action that Barnes earned his DSC was in Ho Nai Village, which sat right across the road and main gate from

Camp Frenzell-Jones, the Brigade Main Base for the 199th LIB. This was also the same action where CPT Robert Tonsetic and SP4 Robert Archibald of C14-12, 1999th LIB carned their C/4-12, 199th LIB earned his Distinguished Service Cross as well).

Beginning at 1600 hours on May 24th, Delta Company, jittery with weapons and gear at the ready, lined up by platoon on the runway. It was lifted en-masse by Chinook from the airstrip at FSB Buttons to FSB Myron, which lay in the middle of hostile territory, 10 kilometers inside the Cambodian border.

In a letter home dated May 28th, Bob Pempsell hurriedly wrote, "Well, I finally made it. I'm inside Cambodia. It seems that this is where the gooks are. There is action all around us, but for the company, we luckily haven't seen any yet. We have, however, found a lot of well-used trails."

For the next two weeks, Delta continued to hump, search and seek in the vast wilderness around FSB Myron, their luck still with them as they still had not engaged in any major firefights with the NVA. There was, however, a lot of sniper and harassing fire from the still unseen NVA who were quietly shadowing their movements. (Despite this relative lack of activity, Delta had been alongside of Bravo Company on June 1st and the 11th when they found the huge NVA depot and bunker complex. See Chapter 3).

On June 16th, the company, drenched with sweat, was busy busting through a stretch of particularly thick jungle and rocky terrain two miles east of FSB Myron. They were searching for a suspicious spot in the jungle that had been seen from the air by LTC DeLeuil's C&C chopper.

At the point and leading the company through this green-hell was SGT Bob Prinn, Sp4 Roger Levins and their Labrador Retriever from the 199th's attached, 76th Combat Tracker Team.

(The 76th CTT was formed at Camp Frenzell-Jones in April, 1968. In contrast to the 199th's other scout dog unit, the 49th Infantry Platoon Scout Dog, the 76th used Labrador Retrievers instead of German Shepherds. The highly-trained dogs and their finely-tuned handlers were trained as trackers and searched for signs of the enemy through such avenues as trails, footprints and broken twigs or disturbed vegetation. Their main weapon, however, was the dog's keen sense of smell. Four members of the 76th Combat Tracker Team were killed in action while serving with the 199th Infantry Brigade in Vietnam/Cambodia from 1968-1970).

At 1110 hours, the dog sensed danger and abruptly alerted. Levins and Prinn immediately halted the column as five NVA soldiers suddenly appeared, carrying AK-47's at the ready and wearing light green uniforms, passed by their position to within 30 meters.

Prinn was immediately on the horn, relaying what they had just seen to CPT Barnes, who was then several meters behind them with the main body.

A fire mission was called in and within minutes, Delta Battery at Myron was cranking out 105mm rounds into the area in front.

For nearly 20 ground-shaking minutes, the artillery kept raining down steel to Delta's front and flanks. Several secondary explosions were heard during the barrage, indicating that something worth investigating was in their immediate vicinity.

At 1200, the shelling stopped and CPT Barnes subsequently sent out a small, five man recon patrol which included the two tracker team members, the dog and three other line grunts. They moved out in the direction where the NVA were last seen.

The patrol, led by Sp4 Levins and his dog, crept as quietly as they could past the uprooted trees and destroyed vegetation caused by the 105mm rounds. The smell of cordite and gunpowder still hung lazily in the air. As they glided forward to a small, open area that was not as dense as the surrounding jungle, the Lab again jerked to a stop and alerted. Sp4 Levins fixed his gaze a few feet ahead of them, trying in vain to see what the dog was staring at.

"BBBBOOOOMMMM!!!   BRRRRRRPPPPPPPPPP!!!! SWWWWOOOOOSSSHHH!!!" The quiet was pierced by the deep, deadly detonation of a Chicom Claymore Mine strung up in the trees at head level. Murderous RPG and machine gun fire then engulfed the five men. Four of them were hit instantly. They fired back what rounds they could muster from their CAR-15's and M16's before the NVA quickly withdrew deeper into the bush at approximately 1210, leaving the small patrol bleeding and groaning on the putrid jungle floor.

PFC Victor Gianfala, PFC Larry T. King and SGT Tony Prinn were bleeding profusely from innumerable shrapnel and bullet wounds. Sp4

Richard Levins, 25 and married was killed in seconds, absorbing most of the ball-bearings in his body from the mine. According to those that witnessed the short firefight, Levins went down firing his CAR-15 on full-auto.

By 0700 on June 18th, Delta was up and moving, busily breaking apart their night defensive position, policing up their trash and wearily preparing for another days walk. It was already in the low 90's by the time the company moved out and it was the start of their eighth straight day in the field.

The 1st Platoon was on point, followed in the air by LTC DeLeuil in his ever-present C&C chopper. From above, DeLeuil had seen some suspicious looking structures in Delta's AO and he wanted CPT Barnes to check them out as soon as possible.

By 0900, the point squad had broken through the grasping clutches of the thorns and wait-a-minute vines and reached their objective. Surrounded by dying, brown vegetation were three bamboo hootches, approximately 18 inches above the ground sitting on stilts.

CPT Barnes was immediately called to the front as the men formed a tight perimeter around the hootches. There was no talking as the men carefully scanned their surroundings. They then searched each one of the hootches, carefully probing the ground and structures for booby-traps.

Two of the three bamboo structures were empty. However, the third and final one, which lay collapsed upon itself in the tree-line, yielded 32 crates of brand-new, Chinese made SKS rifles. Each crate contained 21 individual weapons. The men in Delta were astounded. In all, 384 rifles were found and recorded.

John Hart of the 1st Platoon remembers, "Those rifles we found were brand-spanking new, still covered with cosmoline grease. Everybody in the company was given one to take home at the end of their tour. We were all very excited."

According to Bob Pempsell, "It was the largest, single cache of SKS rifles found in Cambodia up to that point in the Incursion. Those that were not claimed by company personnel were either sent back to the rear or to the Cambodian Army."

When the word got back to the TOC at Myron about the find, the air over Delta was literally buzzing with activity. Everybody wanted a souvenir to take home. An LZ was cut near the find and Huey helicopters crisscrossed the sky for hours, jockeying for position to come in and land, dropping off warm food and water in exchange for crates of SKS's.

On one of the re-supply choppers was a reporter from the Associated Press. Dressed like he was on an African safari, the newsman stepped off the skids and quickly began interviewing some of the tired yet alert, grungy troopers who were cautiously relaxing on the fringes of the LZ.

According to the reporter, "Though they were no longer on the move after eight days in the field, there were no smiles at the cache site. Soldiers could be seen stationed in the growth around the landing zone. Eyes searched in many directions. Talking was in whispers. Sounds of the jungle came crashing in."

One group of soldiers was asked how they feel when they find a cache.

"Scared," they answered almost in unison.

"Because of the booby traps?"

"No…because a cache means there are people around."

Despite firing on three enemy soldiers who were detected moving down a nearby trail that paralleled the hootches, the rest of the day passed by rather uneventfully for Delta Company.

That evening, the grunts began their nightly ritual of clearing away the bamboo and underbrush and setting up another night defensive position.

No matter how tired or how miserable the conditions were, the standard operating procedure was always the same and always taken seriously. Fighting positions or bunkers were constructed with care, so that two or three men could be in the same position during the night.

Claymore mines and trip flares were set out and carefully hidden 30 meters or so beyond the perimeter, while fields of fire were cut, positions were radioed into the CP and back to battalion, artillery was pre-plotted, weapons and equipment were cleaned, cold C-rat suppers were eaten, hurried letters were written home, guards were posted and the men on LP/OP were sent out several meters beyond the wire, while most of the men prayed that nothing would happen during the next several hours.

Bob Pempsell remembers, "Being mainly with the CP group, I never dug fox holes much. Usually, I would lay on the ground, dog tired, until some critter would start messing with me and wake me up. We did not move around at night because most times, it was so dark in the jungle you couldn't see. It was a mess trying to find the next guy who was on radio watch. If the situation was right, I would sometimes string up a hammock between two trees and sleep there."

After all the air and ground activity that had taken place around Delta's position at the cache site during the day, it was no surprise that the NVA knew exactly where Delta Company was located. Shortly after nightfall, the LP/OP's began calling back to the CP stating that they could both hear and see movement out in the jungle. The fear and energy level began to rise.

By the early morning hours on the 19th, a few light ground probes had occurred around the perimeter, mixed in with random but inaccurate sniper fire. The already sleep-deprived grunts were kept awake and on edge. Something was about to happen. Everyone there could sense it...

June 19th, 1970. In the brief minutes before the sun's rays crept into the already alive jungle in Southeast Asia's false-light, Delta Company carefully observed the customary stand-to.

For several minutes, each individual wearily scanned the jungle before his position, making sure there were no enemy soldiers moving about before the GI's crawled out of their bunkers and began preparing breakfast and packing their gear for another day in their tour of duty.

"On that day, we didn't move out at first light like we usually did," says John Hart. "To break up the routine in case there were any NVA in the vicinity watching, which there was, we began to move out by 0900 or so. We thought that it would be another day of monotonous patrolling. Little did we know what was in store for us. Our luck had finally run out."

By 0915, after barely inching forward, the point squad found one vacant enemy fighting position, some blue commo wire that disappeared into the jungle and one fresh blood trail going in the same direction as their line of march. The prospect of the day being routine quickly diminished.

"My squad was on point," continues John Hart. "Johnny Watson, my best friend from Mobile, Alabama, was carrying the M60 in line before me. I was carrying the M79 grenade launcher with a .45 pistol and several belts of M60 ammunition. In front of me and Watson was SGT Michael W. Notermann, PFC Ronald R. Stewart and a Kit Carson Scout. While we cautiously probed forward, we suddenly noticed several high-speed trails with recent use intersecting our line of march. They were criss-crossing everywhere. It was really scary. We knew we were in the enemy's backyard."

After breaking through to a recently-constructed engineer road that was made several days prior to back haul supplies found at another cache site, the company continued on its northward azimuth, up one hill and then down another, hoping to reach their daily objective on time and on schedule.

The time was 1202 p.m. As PFC Ronald Stewart and the rest of the men on point rounded a slight bend into an open area on the engineer road, an estimated 20-30 NVA, carefully hidden and camouflaged in new bunkers with overhead protection, cut loose from both flanks on the terribly exposed GI's. It was yet another devastating U-shaped ambush for the men of the 5-12th Infantry.

AK-47's and SKS carbines popped and cracked while RPD machine guns barked and chattered. Chinese Claymores thundered with horrible effect and RPG's whistled and screeched into the men with tremendous impact, spewing hot pieces of burning shrapnel into unprotected flesh and bone.

Reeling and caught off guard, those members of Delta that were not hit in the opening fusillade tried to counter with what fire they could put out. For several minutes, all was total confusion as the NVA achieved fire superiority from the start, raking the kill-zone back and forth.

"The NVA were just waiting on us," exclaims John Hart. "The dinks were even up in the trees firing down on us. It was absolutely terrible."

Hart continues. "We were caught in the open and the NVA took full advantage of that. It seemed as if the NVA were everywhere and when not trying to fire my weapon, I caught glimpses of them darting from tree to tree, firing as they moved."

Sadly, PFC John "Johnny" M. Watson, 20, SGT Michael W. Notermann, 20 and PFC Ronald R. Stewart, 20, were killed instantly, most-likely by the claymores strung up in the trees in the first few seconds of the ambush. They never had a chance to make it out of the line of fire.

Further down the column, Sp4 David M. Drews, Sp4 Rogelio Pena, PFC Ernest W. Currie, PFC Daniel Smith, SGT Robert W. VanWinkle and PFC Stephen L. Hogg were all hit and seriously wounded by the ear-splitting and accurate enemy small arms and RPG fire.

Calls for help were immediately made and helicopter medevacs were on station within minutes. Medevac 15, hovering overhead with its jungle penetrator lowered and ready for pickup, was driven off twice by heavy ground fire. On the third attempt, despite rounds still popping and dancing around and through the unarmed aircraft, it was able to land and get out the most critical of the wounded. Quickly thrown on board, they were taken back to Song Be.

Bob Pempsell remembers, "When the ambush was sprung, one guy was shot by an AK not fifteen meters from me. Those dink bastards just would not give up. Throughout the day, contact was broken and reestablished with the NVA more than three times. They wanted to wipe us out. It got so bad and so intense, I had to start firing my M16 instead of stay on the radio and direct the artillery fire and air strikes. I had mortars blasting away at the dinks a hundred, fifty meters out and also had Blue Max gunships on station, which were tearing up everything with their mini-guns twenty meters outside of our perimeter."

"We were dead in their sights, right out in the open, scrambling to find what scant cover was available," remembers John Hart. "As the firefight progressed in ferocity and numbers, we formed a makeshift perimeter. We were being hit on three sides. I vividly remember one of the medics sitting up, in the midst of the terrible enemy fire, working feverishly on the guys that had been hit. He was screaming at the top of his lungs at anything and everything. The whole scene was just surreal. It really cannot be described."

After nearly four hours of heavy and sustained firing, the company finally broke contact for the last time and withdrew several meters down the road. As air and artillery continued to pound the enemy positions throughout the rest of the day, Delta Company settled in for another long night in Cambodia.

John Hart emotionally reflects on the events of that day.

"I had so many good friends over there, but Johnny Watson stands out to this day. I can't explain why. He was as black as coal and his personality and smile were as white as could be. Johnny could literally light up a room with those traits. I was just 20 years old and he was a little older than I was, married with two kids. He was like a big brother and a mentor to me and thus, I have a special, unending love for him. Although he brought a lot of laughter to me and those that knew him, I have shed more tears over his loss. Despite our age today, our fallen comrades and brothers have not aged at all. I remember everyone being young."

Previously that afternoon, while Delta Company was heavily engaged, Bravo Company, rapidly moving towards the sound of the guns to reinforce them, was also in contact less than a mile to their northwest.

Jim Horine of Bravo Company explains, "The closer we got to Delta, the heavier the firing became. We too started taking incoming fire. There was a large hill to our front and by the time we got to the base of it, the NVA were firing on us incessantly. We literally had to low-crawl up the side of this steep hill to make it to the top. By the time we got up there, we were dog tired. The CO ordered us stop and maintain our positions."

That evening, while the men were setting up their own NDP, an NVA soldier, crawling unseen towards them, suddenly stood up on one knee and fired a round from his RPG into the perimeter. Sp4 Sydney Flame and PFC Raymond F. Kleinhenz were both hit by the projectile.

Shortly thereafter, Charlie Company, operating several klicks from the fighting, wandered into a huge Communist training area that was more than 500 meters in diameter. Over 40 hootches were counted, along with 22 bunkers, two obstacle courses with a mock American perimeter and TOC. Two classrooms with bleachers that could seat 200 personnel each were seen, along with various training aids, including wooden silhouettes of GI's and various Chinese and American mines and RPG's.

The dark, eerie blackness of night engulfed the shaken survivors of Delta Company as the last of the sun's rays drifted lazily out of view. Bloodied and battered, the survivors licked their wounds and hunkered down in their freshly-dug foxholes and bunkers.

Contact with the enemy resumed on and off throughout the night at brief intervals. Like unseen phantoms, the NVA probed all up and down Delta's line. Never attacking in force, the NVA instead worked in three-man cells to snipe at and harass the Redcatchers at different points along the line. They also banged on the surrounding trees and bamboo, trying to get the GI's to give away their positions with automatic weapons fire so they could chop them up with a well-aimed RPG round.

Shortly before midnight, while speaking on the radio with LTC DeLeuil, CPT Barnes, under heavy stress, broke down and asked to be relieved from command. Suffering from stress and combat fatigue, he had just seen too much in his two years in-country.

Surprised at CPT Barnes's request, DeLeuil ordered him to sit tight, he would come and get him out in the morning.

The news did nothing to calm the nerves of the enlisted men and junior officers in the company.

"When I get out of this f---ing place, I don't ever want to think of it again and I don't want anything to do with the US Army, anything that is green or something that goes f---ing bang. Ever! How can the rest of our lives ever compare to what we are doing now Sergeant Kemp, and how in the hell can you possibly cope with this shit we are going through without going absolutely crazy? I mean, I have known you for what, three months now and during that time, you never seem to worry about anything or get really upset. Shit, you don't even use curse words. Who the hell do you think you are, Superman or something? What's your secret?"

"Well Private Grant, let me explain it to you. It's really very simple. My faith in Jesus Christ and the Bible gets me through all of this. It gives me the strength and the quiet peace of mind that I need in order to survive this place and this whole experience. According to what God's word says in the Bible, its not over when we die. The best is yet to come."

"And that brings you hope and takes away all of your concerns and worries then, SGT Kemp? No way man, that's just way too easy and what about being a soldier and killing people? What does your Bible say about that? It just not that easy."

"Yes, Private Grant, it is that easy. It's called faith in the risen Savior that died for all of us, and I do believe in everything that the Bible says. And to answer your question about being a soldier and killing. The Bible mentions the profession of being a soldier in high regard. Several soldiers in the Bible are mentioned, such as Cornelius in the book of Acts and the Roman Centurion in Matthew 8. Being a soldier was and still is consistent with what we Christians call a Biblical worldview. As for fighting and killing your enemy, there is no condemnation in God's eyes for the soldier that does his duty with dignity and honor during wartime. There is a difference between doing your duty and going home and liking the evils of war."

"So Sergeant Kemp, it's just that simple."

"That's right, Grant. It is. Before we get the word to ruck up and move out yet again, let me leave you with this. It is from the 91st Psalms and it is often referred to as the Soldiers Prayer."

"We live within the shadow of the Almighty, sheltered by the God who is above all gods. This I declare, that he alone is my refuge, my place of safety; he is my God, and I am trusting him. For he rescues you from every trap and protects you from the fatal plague. He will shield you with his wings! They will shelter you. His faithful promises are your armor. Now you don't need to be afraid of the dark any more, nor fear the dangers of the day; nor dread the plagues of darkness, nor disasters in the morning. Though a thousand fall at my side, though ten thousand are dying around me, the evil will not touch me. I will see how the wicked are punished, but I will not share it. For Jehovah is my refuge! I choose the God above all gods to shelter me. How then can evil overtake me or any plague come near? For he orders his angels to protect you wherever you go. They will steady you with their hands to keep you from stumbling against the rocks on the trail. You can safely meet a lion or step on poisonous snakes, yes, even trample them beneath your feet! For the Lord says, "Because he loves me, I will rescue him; I will make him great because he trusts in my name. When he calls on me, I will answer; I will be with him in trouble and rescue him and honor him. I will satisfy him with a full life and give him my salvation."

The deep whomp, whomp, whomp of LTC DeLeuil's inbound Huey was heard long before it flared into Delta's perimeter at 0845 the morning of June 20th. The men were still uptight despite a relatively quiet night and now they were getting a new commander at this critical time.

Within seconds of the aircraft touching the ground, CPT Barnes was on and Delta's new commanding officer, CPT Mack Gwinn was off. Bob Pempsell was huddled around the CP when the new officer jumped off the skids.

"When CPT Gwinn first walked towards us, all I could think of was John Wayne. He was wearing camouflage fatigues with grenades and a huge knife strapped to him. Knowing that first impressions are not always what they seem to be, Gwinn ended up being a really good leader. At this time, however, he still had to prove himself."

As DeLeuil and Barnes were just about to clear the tree line, the NVA, hidden on a rise to their immediate front, opened fire with a crew-served weapon, both on the helicopter and the grunts on the ground. The Huey was hit and shuddered, barely gaining enough altitude to limp back to FSB Myron. (When they landed at the firebase, there were several slugs inside the craft's engine compartment). Bob Pempsell was once again on the horn, calling in artillery and gun-ships on the NVA.

Refusing to fall back under the heavy American firepower, the Communist soldiers crept as close as possible to Delta's perimeter so as to "hug their belt." This technique was often done by the NVA, having been learned early in the war, so as to negate the use of allied air and artillery support and to increase the possibility of inflicting friendly-fire the closer one got to the Americans. It took balls of steel to close within spitting distance of your enemy.

To keep from being hit by their own supporting fire, Delta began pulling back even further down the engineer road it had wearily trudged up the day before the ambush. Incredibly, the NVA still continued to shadow the company's movements, taking pot shots at the Redcatchers as they re-deployed.

As the air and artillery fire were walked in danger close, with shrapnel whizzing by both combatants, the NVA finally stopped their advance and broke contact. They withdrew back to the hill from where they had come from. Finally, all was quiet on the battlefield.

Both sides now came to a complete stop, nervously waiting and watching to see who would make the next move. "The air was so thick with fear and dread, you could have cut it with a knife," remembers one infantryman.

After two days of fighting, it was obvious that Delta Company was operating in a super-hot area. Throughout the rest of the day, the RTO's and officers listened in astonishment as North Vietnamese radio traffic was broadcast through their own radio frequencies. They were making no effort to conceal their intentions or their conversations.

Just what the NVA were fighting so hard to keep from the Americans from was yet to be seen. It was widely known to both sides that per President Nixon's orders, the United States Army and all of its components had only 10 days left in Cambodia. The NVA, as the men of D/5-12 were finding out, were going to make those last days as rough and as deadly as possible.

Beginning their tenth day in hostile territory, Delta Company was tired, haggard, filthy, jumpy and nervous to no end. Only those men who have been in this type of situation can best describe the emotions and mind-set as to what it really like. In short, it was and is hell on earth.

Uniforms were rotten and in tatters. Physically and emotionally, the men were drained and many were showing the early signs of malaria.

Twenty and twenty-one year old men gazed at one another and their surroundings through glossy, far-away eyes that were black and hollow. Called the "thousand yard stare," the look came from young men that had just seen too much.

Most of the individuals in the company, like the North Vietnamese opposing them, had fathers or uncles and cousins that had fought against the Japanese or Germans during World War II.

This, however, was a type of war being waged that their relatives would not recognize or truly understand. From 1941-1945, the average age of the American foot soldier was 26. The average age from 1965-1975 was 21.5 years old.

After spending a couple of weeks or a month at the front, soldiers during WWII, along with their entire unit, were rotated back to the rear, far behind the lines, for weeks or months of rest, training and recuperation. Before the mammoth Normandy Invasion of June, 1944, some soldiers trained for nearly two years in England and Scotland for the final act.

Besides a chance of enjoying a one-week R&R respite from Vietnam in Japan, Thailand, Australia or Hawaii, the soldiers in Vietnam and Cambodia had no such luxury of getting away from the fighting and getting shot at, rocketed or mortared.

Death or dismemberment was always present in both the cities and jungles of Vietnam in a conflict that had no boundaries, front lines or secure areas. During a soldiers 365 day tour of duty in Vietnam or Cambodia, the individual that possessed a combat MOS (Military Occupational Specialty) had a better than average chance of seeing or being exposed to at least 240 days of some type of combat action. In WWII, the combat soldier, depending on the campaign, would go for weeks or months without hearing a shot fired in anger.

"The day after CPT Gwinn took over (June 21), we again spent the day on patrol. I don't remember how far we walked. It became a habit to count the hills and mountains we climbed instead of the distance," remembers John Hart. "With all the ammunition, food and water we were humping, it became a burden to put one foot in front of the other."

"A hump could be good if the day was cool, the duration brief, or the day netted some enemy kills. Humps were bad if they were long, hot or costly in American casualties. In Vietnam and Cambodia, a patrol that covered a few thousand meters could mean progress measured by the swing of a machete and a day spent leaning back against the elephant grass by passing mortars and machine guns hand to hand up slopes on which a soldier could barely crawl."

By June 22nd, there was a glimmer of hope for Delta. If all went to plan, in two more days, they would be out of hell and back at FSB Myron for a brief respite from the war and in five more days, they would be out of Cambodia altogether.

With renewed vigor at the prospect of getting out of the jungle, the company began moving towards a suitable landing zone for extraction. Despite the recent contacts and heavy casualties taken as a result of the June 16th and 19th ambushes, the men were in good spirits.

By 1110 that morning, the optimistic attitude among the rank and file quickly diminished; the point squad blindly stumbled upon a very recently vacated NVA way-station. The pucker factor was at an all-time high as the men could literally feel that they were being watched from a short distance away.

While investigating five, carefully camouflaged hootches that comprised the site, forty-seven boxes of 75mm recoilless rounds, twenty-five SKS and twenty-two AK-47 crates were strewn about in the open. All were empty. A five-foot wide, high-speed trail ran in a southwardly direction away from the site. This trail in turn fed into a huge, well-kept road that was previously unknown to exist. The road was as big as any seen back in Vietnam. Bad vibes and premonitions immediately began filtering through the company.

Quickly moving out of the area, CPT Gwinn halted the company on a small knoll some five hundred meters from the cache and road. John Hart remembers that the knoll the company stopped on, "Was in a very rocky and hilly area that overlooked a gradually descending ravine. This landmark penetrated further into the Cambodian frontier like a huge snake. The scene was picture perfect and the afternoon sunshine seemed to radiate from the jungle vegetation ahead of us."

Thinking that the rise would make a natural defensive position, CPT Gwinn ordered the three platoons in the company to drop their rucks and begin making preparations for an overnight position.

While the 3rd Platoon secured the site, the 1st and 2nd Platoons conducted short cloverleaf patrols surrounding the knoll.

Bob Pempsell was with CPT Gwinn and the 1st Platoon when they started to maneuver down through the middle of the ravine in their front. "We left one platoon with our heavy gear on top and moved down the valley. Before going a few yards, we noticed yet another trail that ran parallel to our route of march. There were some large hills on our right as we descended further down the valley. Some of my friends told me later that it was another classic North Vietnamese ambush that we again walked into. The time was around 1500 hours."

As in the ambush that had occurred two days prior on the 19th, the North Vietnamese small arms, automatic weapons and RPG fire was beyond description. The bullets and shrapnel tore into the slow-moving column of GI's like a farmer's scythe cutting down grain. Bodies were mangled, bones were shattered and more young men were killed.

"The dinks were in front and on both of our flanks. I actually felt an enemy round whiz by within centimeters of my face. That same round hit and killed CPL Raul DeJesus from New Mexico. He was less than five meters behind me."

The two point men, CPL Allen E. Oatney, 20 and PFC Charles C. Cisneros, 20, were also killed in the opening volley. SSG Donald J. McDonald, Sp4 Paul F. Konno, Sp4 Robert Bekeshka, PFC Richard Lundein and Sp4 Bruce Zehentner were all wounded.

Because the enemy fire was so intense and unrelenting, nobody could make it to where the bodies of Oatney and Cisneros lay. As hard a decision as it was to make and deciding not to risk anymore of his soldiers' lives, CPT Gwinn ordered what was left of the 1st Platoon to fall back and regroup on the knoll where the rest of the company was hunkered down and waiting. Oatney and Rosa were reported as Missing In Action and left where they had fallen.

"The incoming fire from the NVA was the heaviest we had seen, even more so than on June 19th. After I stopped praying in the first few seconds of the firefight, I saw a silhouette behind a bush on the slope of a hill to my right. I fired a full-magazine on full auto at the bush and the firing stopped from that direction," states Bob Pempsell.

He continues. "Screaming CONTACT, CONTACT, CONTACT into my radio's handset, I had fire missions from both gunships and artillery fire on station. I called in the rocket and artillery fire to our north, west and east. They almost had us surrounded. While this was going on, there was so much noise around me, I could barely hear myself, much less the guy back at the TOC on the radio. I used three tubes of 105's, bringing in the rounds closer and closer until they were impacting within 50 meters of our position. The guys back at FSB Myron actually requested that I give them CPT Gwinn's initials so that I could drop it in closer. I also used a pair of Blue Max Cobra's that shot their rockets to within 20-30 meters from us. They were just magnificent. The 2-40th Artillery's commanding officer was orbiting above us during the firefight and he told me to also use 155's and 8 inch guns from the 1st Cavalry as a blocking force to seal off any avenues of NVA escape or reinforcement. I didn't know where in the hell all that stuff was landing and I was more concerned of killing some of my own guys instead of the NVA killing me."

Because of the equally intense and lethal air and artillery fire called in by Pempsell, the survivors of the 1st Platoon were able to beat feet back to a safer distance and rejoin the 2nd and 3rd Platoons that had been listening incredulously to the sounds of the fighting.

"We left two guys back at the ambush site," laments John Hart. "The enemy fire was just to heavy to retrieve their bodies and take them back with us. They were left out there on the trail that night."

After linking up with the rest of the company, Pempsell and CPT Gwinn continued to call in fire mission after fire mission on the now pinned-down North Vietnamese. Despite being hit hard, Delta Company had now turned the tables on the attackers.

Pempsell states, "I remember vividly talking to the pilot that flew on one of the fast movers. Before screaming in over our position, he told us to find as much cover as possible as this would be a big one. After releasing his ordnance, the concussion from his exploding bombs literally lifted us off the ground for a second or two. When he was finished, he did a couple of victory rolls as he flew over us. That helped to cheer us up a bit."

By early evening, Bravo Company, sweating profusely and of breath from trying to reach their brothers in Delta as quickly as possible, broke through the underbrush to the knoll and collapsed, exhausted. Now reinforced with another line company, CPT Gwinn and the rest of the grunts in "Dying Delta" as the company came to be called, were eager for some payback and hoped that the NVA would oblige them by attacking their easily defensible position. The night passed uneventfully.

Priority number one for Delta and Bravo Companies on the morning of June 23rd, two days before the battalion was scheduled to leave Cambodia, was to recover the bodies of Oatney and Cisneros, who were still listed as MIA.

After a nerve-wracking journey back down the ravine, CPT Gwinn and the rescue team found Oatney and Rosa lying in the same spot where the ambush was first sprung. As they cautiously crept forward, there was a quick burst of fire from an enemy AK-47. A round actually hit CPT Gwinn's wristwatch and blew it from his forearm. Returning fire, the small patrol hit the dirt and waited to see what would happen for several tense minutes. Thankfully, nothing else materialized.

Since the NVA were still around in force, the search party gingerly but quickly searched the bodies for booby-traps by using a rope and then hastily carried them back to the company perimeter. The North Vietnamese let the GI's carry back their dead unmolested.

To say that the last twelve days for Delta Company (and the rest of the battalion for that matter) were terrible would be a huge understatement.

Seven men from the company (counting Fred Levins from the 76th CTT) had been killed in less than a week and over forty became casualties from either hostile enemy fire, symptoms of malaria or from other forms of illness. Cambodia was taking a toll on not only Delta Company, but the rest of the battalion as well. As CPT Gwinn and the rest of his soldiers were extracted by Huey back to FSB Myron late in the afternoon on June 23rd, they couldn't help but think about what a bitch the whole experience had been.

In addition to Delta, Alpha, Bravo, Charlie and Echo Companies also began filtering back into FSB Myron from the dark clutches of the jungle on the 23rd and 24th of June, in anticipation for the battalion's redeployment back to Vietnam which was slated to be completed by the evening of June 25th.

While the infantrymen were busy coming in, the never sleeping, never resting artillerymen from Delta Battery, 2-40th Artillery were working frantically to tear the basecamp apart. Nothing was to be left behind and the site was to look as though nothing had ever been there.

Beginning at 0900 on the morning of June 25th, Chinooks from the 1st Cavalry Division began hovering over FSB Myron like huge praying mantises, waiting to swoop in and pick up the men of the 5th Battalion, 12th Infantry and take them from one war zone to another. Bravo Company, the last unit to board the aircraft and fly back to Vietnam, had to wade through a literal sea of reporters and photographers as they left Cambodia for good.

That night, after making stops at Song Be and Tan Sanh Nhout Airbase, the last remnants of the Warrior battalion slogged into Camp Frenzell-Jones for a much-needed week-long stand-down. After taps was reverently played across the post, the Beatles song "Let It Be," was heard quietly and emotionally resounding through the company streets. All that remains today of the Cambodian Incursion for the survivors of the 5th Battalion, 12th Infantry, 199th Light Infantry Brigade that fought there are the vivid memories and the terrible nightmares...

Epilogue

THE WARRIOR

(April 7th, 1968-October 15th, 1970)

Arriving complete and en-masse on Continental 707's from Ft. Lewis, Washington, at Bien Hoa airbase on April 7, 1968, the 5th Battalion, 12th Infantry "Warriors" gave the 199th Light Infantry Brigade a fourth infantry battalion. The assimilation of 5-12 gave the 199th LIB even greater flexibility and movement in field operations, thus living up to the Brigade motto, "Light, Swift and Accurate." (The Warrior battalion was also assigned to the 199th LIB because of the heavy losses in killed and wounded that the unit had sustained during the Tet Offensive fighting of February, 1968).

The 12th Infantry Regiment is one of the oldest and most-historic organizations in the United States Army. Formed in early July of 1798 in response to rising tensions between the newly created United States, Britain and France over trans-Atlantic shipping rights, the 12th Infantry was quickly disbanded less than a year after its creation.

When the fighting in the War of 1812 began, the 12th Infantry, along with several other regiments, was again activated. Seeing little action or combat against the British until September of 1814, the 12th Infantry was on duty at Ft. McHenry in Baltimore Harbor, Maryland, when the British ferociously attacked.

On the night of September 13, 1814, British warships unleashed a horrendous and colorful cannonade against the beleaguered fortress and its defenders. Held captive on one of the British warships was a young American named Francis Scott Key. It was the colors of the 12th Infantry Regiment and the flag of the United States flying at Ft. McHenry that Key was looking at when he composed the National Anthem.

More than thirty years later, during the war with Mexico, the 12th Infantry again saw fierce action, fighting gallantly in battles against the Mexican Army at National Ridge, Paso de Overjas, Plan de Rio and Coutreras.

From April 1861 to April 1865, the 12th Infantry Regiment was a part of the bloodied and battered Union Army of the Potomac. Being a regular Army unit, the 12th Infantry saw extensive and heavy fighting in some of American history's bitterest and costliest battles.

During the Peninsular Campaign of June 1862 near Richmond, Virginia, the regiment fought in its first battle against other Americans led by General Robert E. Lee at Gaine's Mill. The regiment suffered more than fifty percent casualties in killed and wounded after taking the lead in a charge against deadly Confederate Enfield rifles and 12-pound artillery cannons crammed with canister rounds. Replenishing its ranks with fresh troops in the months after their baptism of fire, the 12th Regiment went on to fight stoically in other major engagements of the War Between the States at Antietam, Fredericksburg, Chancellorsville, Gettysburg, Spotsylvania, Cold Harbor and the Wilderness.

During the Siege of Petersburg, Virginia, in March, 1865, the regiment again suffered overwhelming casualties after assaulting the Confederate breastworks that surrounded the city. This action would be its last of the war before it was sent back to Washington D. C. for guard duty.

In the era known as the "Old West," the regiment added two Indian Campaigns to its battle streamers, first against the Modoc tribe in California in 1872-1873 and then against the Bannocks in 1878. Portions of the 12th Infantry also took part in guerrilla actions against the Sioux and Lakota people at Pine Ridge, South Dakota, in 1890 and 1891.

(The heraldry and tradition of the 12th Infantry Regiment's Indian service continued well into the 20th Century and Vietnam. Instead of normal phonetic alphabet company designations, such as Headquarters, Alpha, Bravo, Charlie, etc., the 12th Infantry adopted Native American names such as Huron, Apache, Blackfoot, Comanche, Dakota and Erie Companies).

At the very end of the nineteenth century and into the beginning of the twentieth, the 12th Infantry once again found itself at war. During the Spanish-American War, the unit was deployed to Cuba to fight Insurrectos in June of 1898. Its moment of glory came at the Battle of El Caney where the men from the unit stormed the once impregnable Spanish fortress there and captured the prized enemy colors. In February of 1899, the regiment, along with several others that had combat experience in Cuba, was sent to the Philippine Islands to bolster U.S. forces that were engaged in a brutal and savage form of guerrilla warfare. This was to be the first, but not the last time the unit fought in the jungles of Asia.

The fighting there is known to historians as the Philippine Insurrection and the 12th Infantry did not return to the United States from overseas until the summer of 1912. During World War I, the unit was not deployed overseas.

By World War II, the 12th Infantry Regiment was under the command of the 4th Infantry Division and was sent to England in January of 1944 in preparation for the much- awaited invasion of Adolph Hitler's, "Fortress Europe."

On D-Day, June 6th, 1944, the 12th led the amphibious assault of the 4th Infantry Division into the murderous German MG-42 machine-gun and 88mm shellfire on Utah Beach. Throughout the rest of 1944 and 1945, the 12th participated in five combat campaigns throughout Western Europe. The highly prestigious Presidential Unit Citation and the Belgian Fourragere were awarded to the unit for its actions during the largest battle the United States Army has ever fought, the Battle of the Bulge.

After the German surrender, the 12th Infantry, along with the rest of the 4th Infantry Division, returned to the United States in July of 1945. The unit was subsequently deactivated on February 27, 1946 at Camp Butner, North Carolina.

During the Cold War era from 1947 to 1965, the 12th Infantry went through a series of reorganization and training phases until the U.S.

Army's buildup in Southeast Asia and Vietnam. As troop numbers in Vietnam steadily increased in 1965 and 1966, the 1st, 2nd and 3rd Battalions of the 12th Infantry deployed to Vietnam with the 4th Infantry Division from August to October 1966. The 4th Battalion, 12th Infantry was activated and assigned to the 199th LIB in June.

When the 5th Battalion of the 12th Infantry joined its sister battalion in the 199th Infantry in Vietnam in April of 1968, the 12th Infantry became the only unit to have more of its battalions deployed to Vietnam than any other infantry regiment.

Although its history is not as long and illustrious as the 12th Infantry Regiment, the 40th Field Artillery Regiment has served with distinction and honor in the United States Army since being organized and constituted in 1918.

With the motto, "All for One," the 40th Artillery participated in five campaigns in Europe during World War II, pumping out 105mm artillery rounds for General George S. Patton's famous Third Army. Years later, the 2nd Battalion of the 40th Artillery Regiment was one of the first, original units to join the 199th LIB when it was being created at Ft. Benning in June of 1966.

One artillery battery from the regiment was designed to give timely and accurate 105mm fire support to each of the Brigade's four infantry battalions. The term, "Light" in the 199th's, "Light Infantry" designation meant that the heaviest fire support came from the 2-40th Artillery's towed, M-102, 105mm Howitzers. The "Separate" meant that the new Brigade would be a "separate," self-supporting entity in itself and not be under any divisional or similar control. (In Vietnam, despite all the other battalions, brigades and divisions that were deployed to the war zone from 1965-1970, the 199th LIB remained the only separate infantry unit to fight there).

According to CPT Kay Moon, the 5-12th Infantry's S-3 Air and Dakota Company commander from 69-70, "The LT stood for Light and the SEP for Separate and that's just what it meant. If we didn't have it strapped to our backs, we didn't get it most of the time. We also didn't have a lot of internal transportation, so we walked. Any truck or aircraft were mainly a Corps asset."

Delta Battery of the 2-40th Artillery was not created until October 28, 1968. "The first ninety days of existence were filled with sweat and toil, but were none the less fruitful for Delta Battery." In the battery's first five months of operation, the crews had sent out over 30,000 rounds in support of 199th combat operations.

In February of 1969, the battery fired 10,031 rounds alone. By September of 1970, the 2nd Battalion, 40th Artillery had fired over 1,100,000 rounds in support of combat operations. It was also the only artillery unit in the III Corps Tactical Zone that could boast participating in fire missions from inside Cambodia to the South China Sea.

Robert Schwaber was originally assigned to Delta Company of the 2nd Battalion, 3rd Infantry as a rifleman when he arrived in country in February of 1969. After a month of humping as a grunt through the sweltering rice-paddies of the Pineapple region, he was transferred to Delta Battery of 2-40. According to Schwaber, "It was a good transition. I learned very quickly how the artillery worked and functioned because I listened and paid attention to what I was being taught. I worked mainly in the FDC, or the Fire Direction Control center. We plotted and calculated where the rounds were going to hit when fired. We could hit out targets precisely at 9000 meters or more and when the call came in for a fire mission, we were putting rounds downrange and on target in 45 seconds or less."

When Bob Pempsell first joined Delta Company as an artillery sergeant, he carried his M16 rifle, four to five magazines and a few pieces of equipment. By the time the battalion left in late June, he was carrying well over 20, fully-loaded magazines, a couple of ammunition bandoleers for the M60 machine gun, two fragmentation grenades, six smoke grenades, a PRC-25 radio and everything else that he needed to survive.

When the Warrior battalion arrived in Vietnam in April of 1968, it was at a critical juncture for the 199th. The heavy fighting of the Tet Offensive was just coming to an end and the upcoming May Offensive was on the horizon. Surprisingly, the fighting during the May Offensive of 1968 was just as bad, and in some respects, even worse and more intense than Tet. Had it not been for the battalion's timely arrival, casualties within the 199th would have been higher.

By April 15, the battalion had finished its orientation and in-country training and deployed into the field in the nether-regions south and southwest of Saigon. Firebase Choctaw was the first base camp established as the line companies of 5-12 began conducting combat patrols and nightly ambushes. The primary mission was to thwart enemy infiltration and rocket/mortar attacks against the Saigon/Bien Hoa/Long Binh area. The battalion suffered its first two losses on April 18, when two young infantrymen from Charlie Company were killed in action with a Main Force Viet Cong unit.

Bert Ovitt, a four tour veteran of Vietnam, served in the battalion from its stateside training at Ft. Lewis, Washington, until after the Cambodia Incursion.

"When I heard that the unit was being trained for Vietnam, I volunteered to go to the 5-12 under my now secondary MOS (Military Occupational Specialty) of 11B10, Infantryman. Since it wasn't official that they were being deployed at that time, the unit I was with could not block my reassignment. I finally received orders for 5-12 and to report around the 21st of March, 1968. When I arrived, everyone was gone on leave prior to deployment, except for a skeleton crew which consisted of one supply man, one clerk, an NCO and a duty officer. This is when I met Raymond Villarpando. He was being reassigned from Hawaii and was coming into the unit from dog handler duty. He was assigned to 2nd platoon and I was assigned to 3rd platoon. We spent the next seven to ten days together almost constantly, with Ray asking me about Vietnam. We struck up a good friendship. We arrived in Vietnam on the 2nd or 3rd of April, 1968 and spent a week or two doing lots of training and adjusting to the heat. I saw Ray fairly often. My platoon did not assign me to any squad because they had trained together for such a long period of time and setting up their chain of command, they did not want to disrupt it. But, someone felt obligated to put me in a responsible position because it was my second tour in Vietnam. I kept telling them that I had been running convoys and boats and everything else but infantry and the only thing they knew is that I had already finished a tour and that meant a lot to those in charge. The platoon leader finally decided to make me his Radio Telephone Operator (RTO), where I could be close at hand for information. The first days out in the field, I knew enough to stay away from the LT with that radio on my back and the antenna sticking up."

Colonel Albert W. Malone had the honor of commanding the 5-12th Infantry from November 9, 1968, until mid-May, 1969. According to Colonel Malone, "The lessons learned from our operations in the Pineapple region could fill volumes. Perhaps the best lesson of all was that we were extremely well schooled in this type of warfare and if we put into practice exactly what knew rather than blindly following standard infantry small unit tactics, we would do very well. After several weeks of operating in the Pineapple, we killed a courier going from one VC District Committee to another and we found out that we were more effective than we knew. The first District Commander had complained to the other that they had been invaded by U.S. Special Forces! As you can expect, we were quite delighted to find that the 5th Battalion, 12th Infantry was now a 'Special Forces' unit. We made widespread distribution of that VC report throughout the battalion."

On May 13, 1968, Alpha Company, along with a small element from the reconnaissance platoon of Echo Company Recon, 2-3rd Infantry, Delta Company, 2-3rd Infantry and armored personnel carriers from D Troop, 17th Cavalry, became embroiled in a fierce firefight that would see the Medal of Honor posthumously awarded to a young, 5th Battalion Warrior.

Early in the twinkling morning hours of May 12th, a small reconnaissance element from E/2-3rd Infantry was quietly inserted into the swamps and rice paddies southwest of the Binh Dinh Bridge and FSB Hun.

Later that day, the men from the "Old Guard" Recon patrol spotted movement to their direct front. What they saw coming straight at them was a significantly larger enemy force that was composed of NVA soldiers.

Unable to escape detection, the recon patrol detonated their claymore mines and sent a fusillade of M16 and machine gun fire into the ranks of the advancing enemy soldiers. Momentarily caught off guard, the larger enemy force rebounded from the unpleasant surprise and returned heavy and accurate AK-47 and RPG fire on the outnumbered recon patrol.

The team leader and two other recon soldiers were seriously wounded by the enemy's return fire. Despite calling in repeated air and artillery support, the Echo Company Reconnaissance patrol was in danger of being overrun and annihilated. The call went out for an emergency reaction force and Alpha Company, 5-12th Infantry answered the call.

Because the area of contact was in watery and swampy terrain, Alpha Company was forced to board World War II era "Rag Boats" or troop carriers, and chug downstream as fast as the current and the outdated motors could carry them.

Less than 20 minutes after leaving their embarkation point, the rescue column was stitched by AK-47 rounds and blasted by RPG fire from hidden Viet Cong assailants on the riverbank. Alpha Company suffered three, seriously wounded soldiers from the start. Undeterred, the rest of the reaction force disembarked several meters below the trapped and bloodied recon team. Unbeknownst to the officers and grunts of Alpha, they were advancing into a well camouflaged and heavily defended bunker complex belonging to elements of the 9th Viet Cong Main Force Division. (The enemy soldiers from the 9th VC Division had been fanatically fighting in and around Saigon since the start of the May Offensive on May 4).

Taking heavy fire every step of the way, forward elements of Alpha Company finally reached the shaken survivors of Echo Recon, who were desperately seeking protection and cover from behind a small rice paddy dike.

While heavy air and artillery strikes churned the enemy-held nipa-palm grove into small chunks, the Warriors from Alpha deployed into a company skirmish line and assaulted the tree line, all the while laying down a heavy barrage of M16, M60 and M79 fire. It was a charge that would have made any World War I officer envious and proud.

According to an after action report on the firefight, "After the company moved into the area of contact and overran the first line of enemy bunkers, Specialist Kenneth L. Olsen and a fellow soldier moved forward of their platoon to investigate another suspected line of enemy bunkers. As the two men advanced, they came under intense automatic weapons fire from an enemy position ten meters to their front. The

Warriors were immediately pinned down. With complete disregard for his own safety, Specialist Olson raised up, exposing himself to the heavy machine gun fire and hurled a hand grenade into the Viet Cong position. Failing to silence the enemy's fire, he pulled the pin on another grenade and prepared to throw. As Specialist Olson again exposed himself to the insurgents, he was wounded by enemy fire, causing him to drop the activated grenade. Realizing that the grenade would explode at any moment, Specialist Olson, with utter disregard for his own life, threw him-self upon the hand grenade, taking the full brunt of the explosion. As a result of his actions, Olson was the only member of the unit to receive serious injury from the grenade."

Mack Begley served as an artillery reconnaissance sergeant from Delta Battery, 2-40th Artillery with 5-12 during his tour of duty from February, 1968, until March, 1969. On August 5, 1968, he was in the firefight that wounded both the Brigade commander and the battalion commander. (While in Vietnam, the 199th LIB lost one Brigade commander killed in action and two wounded in action, along with a deputy commander).

"On or about August 5, 1968, we were on a search and destroy mission in the Rung Sat Special Zone. We were moving a battery of 105mm howitzers to Nha Be in support of operations there. As we were cruising down the river, I believe in an older Korean-vintage LST, we were suddenly hammered with AK-47 fire and B40 rockets. The boat had a small helipad on the top and General Franklin Davis had just landed and come aboard with his aide when the rockets came whistling in and struck the boat. The shaped charge went clean through both sides of the boat. It even set the mattresses on fire that were below decks. The CO of the 5-12th Infantry, LTC Herbert Ray was hit in the back, taking lots of shrapnel in the buttocks and arms. He would have been killed if he hadn't have been wearing his flak jacket. His left arm was so tore up that I could see the bone where the shrapnel had tore it away. General Davis was also seriously hit, along with his aide and three other sergeants. The .50 calibers on the top deck began firing and I thought that I was hit because of all the hot casings that were falling on me. I was also having trouble hearing because of all the firing and the rocket blast. I never would have thought that the Viet Cong would have fired on a boat such as the one we were riding on, but boy, was

I wrong. The boats had an automatic M79 grenade launcher on top, along with several .50 caliber machine guns, an 81 mm mortar and several .30 caliber and M60 machine guns. When the boat pulled to shore and let the infantry off, the artillery captain and I orbited the fight in a helicopter, directing air and artillery strikes on the enemy. We also went in and picked up some wounded from the firefight below."

Throughout the remainder of 1968 and 1969, the battalion, along with the rest of the 199th LIB, continued to operate in the rancid rice paddies and leech-infested swamps of the Pineapple Plantation and other miserable areas south of Saigon. In this region, more men from the Brigade were lost to booby traps than rifle and machine gun fire.

In June of 1969, the Brigade, moved more than 80 miles north to the mountains, hills and triple-canopy jungle of Long Khanh and Binh Tuy Provinces. It was the start of a new war for the Redcatchers and they would stay there until the unit rotated back to the United States and Ft. Benning in October of 1970.

Instead of Viet Cong guerrillas, the adversary there were uniformed NVA soldiers form the 33rd NVA Regiment and the 274th Main Force Viet Cong Regiment. Most of the Communist soldiers were fresh off the Ho Chi Minh Trail and well armed and well equipped from the supply depots of the Cambodian sanctuaries. Enemy contacts and firefights picked up in intensity and ferocity while in this new AO. Even while the 5-12th Infantry was engaged in heavy fighting in Cambodia, the other battalions in the Brigade were busy making contact as well.

After returning from Cambodia and after its one week stand down at Camp Frenzell-Jones, the battalion was once again sent back out into the field for combat operations for the remainder of its tenure in Vietnam. Deploying around Tanh Linh, which was an old Special Forces A-Team camp in southern Binh Tuy Province, the 5-12th Infantry constructed firebases called Lam, Deeble, Buzzard, Dreamer, Regal, Flower, Mat and Guin.

The war for little-known or remembered jungle hills and trails continued to be fought and with the exception of a heavy, all-night firefight reminiscent of LZ Brown at FSB Guin on August 9th, the battalion never engaged in combat to the extent of what it had been through during the Cambodian Incursion.

On September 1st, 1970 the 199th Light Infantry Brigade received orders from MACV to start standing down its various units for redeployment back to the United States in compliance with Operation Keystone Robin.

Within the following week, both the 5-12th Infantry and Delta Battery, 2nd Battalion, 40th Artillery were put on an inactive status. Those personnel who had less than seven to eight months in country were given orders to join other combat units still fighting in Vietnam. Many of these former Redcatchers that had raided the sanctuary in Cambodia were infused into the 1st Cavalry Division and fought with the units that they had once supported during the Incursion.

Ducti Amore Parriae
Having Been Led By Love Of Country

DAGO 43/72 Valorous Unit Award USAVGO 2264-52/71

The Valorous Unit Award is awarded by direction of the Secretary of the Army to the following named units for extraordinary heroism while engaged in military operations during the periods indicated.

1st CAVALRY DIVISION and its assigned and attached units:
5th Battalion, 12th Infantry 199th Infantry Brigade
Battery D, 2nd Battalion, 40th Artillery 199th Infantry Brigade

The 1ST CAVALRY DIVISION (AIRMOBILE) and its assigned and attached units distinguished themselves through extraordinary in action against North Vietnamese Army and Viet Cong forces in Cambodia and Northern Military Region 3, Republic of Vietnam, during the period 1 May 1970 to 29 June 1970. Assigned the task of seizing base areas and cache depots and interdicting lines of communication occupied by the Central Office South Vietnam Command located in the Fish Hook area of Cambodia, the officers and men of the division brilliantly launched a coordinated armor, mechanized infantry, air cavalry and airmobile infantry assault deep into enemy territory. Demonstrating rare courage, versatility and aggressive determination, unit personnel engaged the enemy in fierce bunker to bunker fighting to drive him from previously untouched sanctuaries. Through their unmatched professionalism and dauntless actions the members of the 1ST CAVALRY DIVISION (AIRMOBILE) and its assigned and attached units effectively neutralized hostile base areas, supply depots and interdicted infiltration routes while inflicting massive casualties upon the enemy at every encounter. As a result of their valiant actions and poise under fire, division personnel contributed immeasurably to the Free World military effort in the Republic of Vietnam. The extraordinary heroism and devotion to duty displayed by the members of the 1ST CAVALRY DIVISION (AIRMOBILE) and its assigned and attached units are in keeping with the highest traditions of the military service and reflect distinct credit upon themselves, their unit and the Armed Forces of the United States.

REPORT DISTRIBUTED TO ALL THAT SERVED IN THE 5-12TH INFANTRY WHILE IN CAMBODIA

(Typed and submitted by Bob Pempsell, D/5-12. This report was distributed to all members of the battalion while on stand-down at Camp Frenzell-Jones, July 1970).

1. The 5th Battalion, 12th Infantry, usually a part of the 199th Light Infantry Brigade, has been under the operational control of the 1st Cavalry Division since the United States Army entered Cambodia. During this period, the 5th Battalion, 12th Infantry has distinguished itself by the discovery and capture of many enemy caches, which contained impressive quantities and varieties of material. These finds are more impressive since several of them form a supply depot. This annex will discuss these caches and their place in the total picture of the Cambodian Operation.

2. In raw figures alone, what the 5-12th Infantry discovered in Cambodia is impressive. Rice: 320 tons. Individual weapons: 449. Crew-served weapons: 4. Small arms ammo: 429,850 rounds. Assorted large caliber ammo: 4,846 rounds.

3. Enough rice was captured to feed 20 NVA companies for an entire year at full rations or 25 companies at reduced rations.

Enough individual weapons were captured to equip three full NVA infantry battalions and such an amount of small arms ammunition was discovered to fire on AK-47 on full automatic for over 72 hours without stopping. There was enough mortar, recoilless rifle and rocket rounds found to launch an estimated 239 attacks. The tactical value to the enemy of this seized material is obvious. A great number of South Vietnamese and American lives have been saved.

4. Besides these items if immediate tactical importance, the 5-12th Infantry also uncovered a wide variety of miscellaneous equipment, in many ways as valuable as the captured weapons, ammunition and rice. Vehicles and automotive supplies are an important group and under this heading. To date, a total of 13 trucks have been captured, to include four, 2 and ½ ton trucks and nine, ½ ton trucks. In addition, many truck parts have been captured, including 27 differentials, 13 wheels, 43 axels, 5 brake drums and other similar items. Tools, such as an hydraulic jack and repair supplies, such as welding rods, have also been seized. Communication equipment is an equally important group of captured material. Of chief importance in this category are four K-62 FM radios. Since no radios of this type have ever been discovered before in either South Vietnam or Cambodia, these K-62's, by showing the extent and capabilities of the NVA signal system, were of great intelligence value. Radio batteries, with a total of 6,682 captured and commo-wire, with a total of 10,000 feet captured, are also major communication finds.

5. Six caches with a center at YU 09415, form a supply depot. Supply depots, such as, "The City" and "Picatinny East" served as transfer points for supplies brought from various points in Cambodia and destined for use in South Vietnam. Usually one or two rear service groups are responsible for a supply depot. The cache complex found by the 5th Battalion, 12th Infantry, consisted of several major rice caches and a communications cache. Accessible to Highway 14 and 131, it was probably serviced by trucks such as those the 5-12th Infantry found a

kilometer away. This depot is in the area of operations of the 86th Rear Service Group and is part of the larger complex of caches and supply depots that form Base Area 351.

6. By aggressively searching and thoroughly exploiting the caches in the area of operations, the 5-12th Infantry has effectively limited the offensive capacity of the enemy. Through its efforts, a significant portion of an enemy base area has been neutralized.

-LETTER FROM A LOVED ONE-
SUE SPENCER, WIDOW OF ARLIE
"PETE" SPENCER
(B/5-12, KILLED IN ACTION, 15 MAY 1970)

To All The Soldiers of the 5th Battalion, 12th Infantry
Then Serving In Cambodia "To The Good Guys"

June 18th, 1970

I want to first thank you all for the most beautiful roses I have ever seen. Your thoughtfullness is most kind. It is so very hard for me to accept the fact that my Pete, SGT Arlie Spencer Jr., is now home and resting so peacefully.

Pete talked or all of you so very much. He felt so good when his men would come to him with their problems. He said that he felt like a "big brother" and loved every minute of it.

I feel as though I know every one of you personally and I think so very much of you, fighting in another country, far away from your loved ones.

The people here have been so very good to us. Pete had more friends than he ever knew he had. There were over 70 flower arrangements brought to the funeral home and hundreds of cards were sent to all of us. There are so many helpful friends taking care of us.

Pete's relatives came from Florida, Kentucky and other states. He had a semi-military funeral with a 21 gun salute and taps. The Captain, Lieutenant and Sergeant who helped us have been very good to all of us. It made me realize that some of the Army is human.

I have nothing but great memories of my great Pete left. Everything is in God's hands now and I'm sure that he is taking good care of him. I pray to him daily that he will allow all of you to return home safely to your loved ones so they won't have to go through anything that we have had to go through.

If there is anything you ever want, please do not hesitate to write. I'll be more than happy to try and help you. If you need someone to write to, just let me know. I know what lonliness is, but I can always find someone to talk to and it's a lot easier than you, being so far away in a strange country.

We all want to be able to help you guys, so just let us know. Packing up care packages is one of our specialties.

Thank you all once again for the kindness you have shown. God bless you all, and maybe soon you all will be coming home safely.

My love to all of you,

Sue, Mrs. Arlie Spencer Jr.

PS- I have enclosed some newspaper articles that were in the city paper.

199ᵀᴴ LIGHT INFANTRY BRIGADE (SEP)(LT) ORDER OF BATTLE, 1966-1970

2nd Battalion 3rd Infantry
3rd Battalion 7th Infantry
4th Battalion 12th Infantry
5th Battalion 12th Infantry
Troop D, 17th Cavalry
Company F, 51st Infantry
Company M, 75th Infantry
2nd Battalion 40th Artillery
7th Support Battalion

Headquarters and Headquarters Company, 199th
HHC MP Combined Reaction Infantry Platoon
71st Infantry Long Range Patrol Detachment
49th Scout Dog Platoon
179th Military Intelligence Detachment
87th Engineer Company
313th Signal Company
152nd Military Police Platoon
76th Infantry Combat Tracker Dog Detachment
44th Military History Detachment
503rd Chemical Detachment
856th Army Security Agency
40th Public Information Office

COMMANDING OFFICERS, 199TH INFANTRY BRIGADE (SEP)(LT)

COL George D. Rehkopf April 66
BG Charles W. Ryder November 66
BG John F. Freund (WIA) March 67
BG Robert C. Forbes September 67
BG Franklin M. Davis, Jr. (WIA) May 68
BG Frederic E. Davison September 68
BG Warren K. Bennett May 69
BG William R. Bond (KIA) November 69
COL Robert W. Selton April 70
COL Joseph E. Collins July 70
LTC George E. Williams September 70

BRIGADE HEADQUARTERS

Long Binh December 66 - March 67
Bien Hoa April 67 - June 67
Long Binh July 67 - February 68
Gao Ho Nai March 68 - June 68
Long Binh July 68 - October 70

NORTH VIETNAMESE ARMY ORDER OF BATTLE FISHHOOK REGION, CAMBODIA, 1 MAY - 30 JUNE, 1970

(Operational Report, 1st Cavalry Division, Airmobile, 31 July 1970)

Dong Nai Regiment Headquarters
 a. K1 Battalion
 b. K2 Battalion
 c. K4 Battalion

Sub-region 5 Headquarters
 a. Thang Loi Battalion

D168 Regiment

D368 Regiment

81st Rear Service Group
84th Rear Service Group
86th Rear Service Group

90th Recovery and Replacement Regiment

K33 Artillery Battalion

5th Viet Cont Main Force Battalion
    a. 174th NVA Infantry Regiment
    b. 275th NVA Infantry Regiment
    c. Z22 Artillery Battalion
    d. Z24 Anti-aircraft Battalion
    e. Z27 Reconnaissance Battalion

70th Rear Service Group

Song Be Infantry Battalion

Z22, 208th Infantry Regiment

A GRUNT'S PRIMER: TERMS AND SLANG COMMONLY USED BY MILITARY PERSONNEL IN VIETNAM/CAMBODIA

ACAV: Armored Cavalry Assault Vehicle used by mechanized and armor units such as the 11th Armored Cavalry Regiment. These were usually M113 armored personnel carriers modified with M60 and .50 caliber machine guns.

AIRMOBILE: Soldiers and material delivered by helicopter.

AIR CAVALRY: Infantrymen carried into battle by helicopters. This term is usually associated with the 1st Air Cavalry Division.

AK-47: The basic infantry weapon carried by the North Vietnamese and Viet Cong soldier. The weapon fired a 7.62x39mm round on either semi or full automatic. It was hard-hitting, very reliable and made a distinct sound when fired.

AO: Area of Operations. This is where a combat unit is operating in the field to find and engage the enemy.

ARC-LIGHT: Code name for a B-52 Strato-fortress air strike. It was absolutely devastating for anything miles around. The plane could carry eighty-four 500 lbs. bombs inside and twenty-four 750 lbs. bombs outside that were mounted on racks. The B-52's had a range of over 3,000 miles.

ARVN: Army of the Republic of Vietnam. The soldiers the US was supposed to be helping.

BASE CAMP: A field headquarters for a combat unit operating in a specific area of operations. The camps would usually contain all of the unit's supporting components.

BATTALION: A unit usually commanded by a Lieutenant Colonel. On paper, the US Army battalion was composed of approximately 700-800 personnel. However, because of wounds, sickness, emergencies and too many other reasons to list, the US Army Infantry battalion in operating in Vietnam and Cambodia was usually smaller. Each battalion had four maneuver companies (Alpha, Bravo, Charlie and Delta) and one support/reconnaissance/mortar company (Echo).

BERM: A built-up area, usually of dirt, that surrounded a base camp's circular perimeter. It offered some protection against incoming enemy rounds.

BIRD: Slang term used to describe most any plane or helicopter.

BMB: Brigade Main Base. Nickname for Camp Frenzell-Jones, the main base for the 199th Light Infantry Brigade while in Vietnam and Cambodia.

BUNKER: A defensive structure built in the field by the already exhausted infantryman, usually around four to five feet deep with overhead cover.

CACHE: Term used to describe a hidden enemy weapons, food or ammunition find.

C & C: Command and Control. C & C oftentimes denoted the commanding officer's Huey helicopter that orbited the sky over a unit's position in the field.

CHICOM: Chinese Communist. The acronym means weapons and equipment made in China.

CHINOOK: The CH-47 Helicopter.

CHOPPER: Slick or helicopter.

CLAYMORE: An anti-personnel mine that was utterly devastating and effective when used against infantry soldiers. The US manufactured claymore mine was a little larger than a 6x9 paperback book and spewed out over 700 steel ball bearings in a 65-degree arc. The Communist mines were larger.

CO: Commanding Officer

COBRA: The AH-16 attack helicopter.

COMPANY: Unit commanded by a captain that consists of three platoons or 120 men. Rarely was an infantry company up to full strength in the field in Vietnam and Cambodia.

CONTACT: In contact, in a firefight, or engaged in battle with the enemy.

CP: Command Post. Where the leaders or commanding officers of a unit are oftentimes located.

CPT: Abbreviation for Captain.

DOC: Slang for medics or medical aid-man.

DUSTOFF/MEDEVAC: Nickname for a medical evacuation helicopter mission.

ELEPHANT GRASS: Six to ten feet tall, razor-sharp grass found throughout Vietnam and Cambodia.

EXTRACTION: Withdrawal of troops from an area by helicopter, truck or boat.

FIRE SUPPORT BASE (FSB): Often named after lost comrades, wives or girlfriends, a firebase was temporary infantry and artillery patrol base found in remote areas throughout Vietnam and Cambodia. Being at a forward firebase was sometimes more dangerous than being out in the field. The two fire bases that the 5-12th Infantry occupied in Cambodia were FSB Brown and FSB Myron.

FIREFIGHT: In contact or in a gun-battle with the enemy.

FIRE MISSION: An artillery mission in which artillery fire is called in on enemy troops or positions.

FO: Forward Observer. This individual, assigned to an artillery unit but detached to the infantry, called in artillery fire missions and air strikes on the enemy.

FRIENDLY FIRE: A euphemism referring to small arms, air or artillery fire that accidentally lands on or hits friendly troops.

GI: Government Issue. Nickname frequently given to US soldiers.

GUNSHIP: An armed attack, Huey or Cobra helicopter that provides fire-support for units in contact.

HOOTCH: Slang term for a house or some sort of dwelling.

HORN: Term that denotes a radio or other communications device.

HQ: Headquarters.

HUEY: UH-1 Huey helicopter. The most famous and widely used helicopter of the Vietnam War. Often called a "slick."

HUMP: To walk around on foot. This was the favorite pastime for the Infantryman and other combat soldiers.

ILLUMINATION: Flares dropped by aircraft or fired from hand, artillery or mortars.

INCOMING: Receiving enemy small arms, rocket or artillery fire.

IN THE FIELD: A forward area, such as being out in the jungle, where the enemy could be found.

KIA: Killed in Action.

LOACH: OH-6A Light observation helicopter.

LOCK AND LOAD: Phrase that means a soldier had a round in the chamber of his weapon, ready to fire.

LZ: Landing zone for helicopters to land and take off from.

M16: The basic infantry weapon of the US soldier in Vietnam and Cambodia. Manufactured by Colt Arms, the M16A1 fired a 5.56mm round on either semi or full automatic.

M60: The most common machine gun carried by US troops during the war. The M60 or "Pig" weighed 23 lbs. and fired a linked 7.62x51mm round at full automatic.

M79: Single-barreled grenade launcher, similar to a fat shotgun, that fired a 40mm grenade.

MACV: Military Assistance Command, Vietnam. The headquarters for all US forces in Vietnam and Cambodia.

MIA: Missing in Action.

NCO: Noncommissioned officer. NCO's were personnel that had earned the rank of Sergeant up to Command Sergeant Major.

NDP: Night defensive position. A defensive position, usually in a circular shape, constructed by an infantry unit when operating in the field.

NVA: North Vietnamese Army or referring to the enemy soldiers.

105: Numerical designation for the 105mm howitzer, the workhorse of field artillery units. The 2nd Battalion, 40th Artillery used these while in Vietnam and Cambodia.

PLATOON: Military unit composed of approximately 45 men, but more often between 20-30. There were three platoons in an infantry company.

POINT MAN: The lead or point soldier in a unit that is operating in the field. The most dangerous position to have while on patrol.

POP SMOKE: To use a smoke grenade to mark a target, location or Landing Zone.

PRC-25/PRC-77: The official names given to the standard infantry field radio. The "prick" 77 was essentially the same as the PRC-25 with the addition of an encryption device for secure communications. The radio was like wearing a huge target on one's back as carrying the radio was a dangerous job to have.

PUCKER FACTOR: Fear factor.

PVT: Abbreviation for Private.

REDCATCHER: Nickname given to the 199th Light Infantry Brigade. The term was derived from the bright red fireball on the shoulder sleeve insignia and meant that the Brigade was going to kill all the communist "Reds" in Vietnam.

ROME PLOW: Bulldozers fitted with huge blades used to clear the jungle terrain.

RPD: Communist machine-gun, comparable to the US M60 machine gun.

RPG: Rocket propelled grenade. Russian and Chinese manufactured shoulder-fired rocket launcher..

R & R: Rest and recreation. One week of vacation taken during a soldiers Tour of Duty in Vietnam.

RTO: Radio Telephone Operator. The soldier that humped the PRC-25/77 radio.

RUCKSACK: The infantry soldier's favorite piece of equipment. It was the backpack that contained everything the soldier needed to survive and often weighed 70 lbs. or more when fully loaded.

SAPPERS: North Vietnamese or Viet Cong demolitions experts.

SGT: Abbreviation for Sergeant.

SHELL: An artillery round or explosive projectile.

SP4: Abbreviation for Specialist 4th Class.

STANDDOWN: A brief period of rest and refitting after coming back to base camp after a patrol or operation.

STARLIGHT: Code-name for a night-vision telescope. It could be mounted on the M14 rifle and used effectively by snipers.

TOC: Tactical operations center. The nerve center for all communications going to and from units in the field.

TRIP WIRE: A thin piece of wire, similar to fishing line, used by both sides for mines and booby-traps.

USARV: United States Army Vietnam.

WIA: Wounded in Action.

WO1: Warrant Officer 1st Class.

BIBLIOGRAPHY/SOURCES CITED

Coleman, J.D. <u>Incursion: From America's Chokehold on the NVA Lifelines to the Sacking of the Cambodian Sanctuaries.</u> New York: St. Martin's Press, 1991.

Dunnigan, James F. and Albert A. Nofi. <u>Dirty Little Secrets of the Vietnam War.</u> New York: St. Martin's Press, 1999.

McDougal, Larry and Bob Stanard. <u>In Country And On Line With The Light Infantry, 1966-1970.</u>

Meadows, Michael V. "Into the Lions Den." Vietnam Magazine (August 2005).

Nolan, Keith W. <u>Into Cambodia.</u> Novato: Presidio Press, 1990.

Olson, James S. editor. <u>Dictionary of the Vietnam War</u>. New York: Greenwood Press, Inc., 1988.

Pempsell, Bob. "Letters Home, May - June, 1970." Delta Company, 5th Battalion, 12th Infantry, 199th LIB.

<u>Redcatcher!</u> Authorized Army Bi-Monthly Publication. Courtesy of the 40th Public Information Detachment, 199th Light Infantry Brigade.

Stanton, Shelby. Michael Casey, Clark Dongan, Dennis Kennedy. <u>The Army At War: The Vietnam Experience.</u> Boston: Boston Publishing Company, 1987.

Starry, Donn A. <u>Mounted Combat in Vietnam</u>. Department of the Army. Washington D.C., 1989.

Tolson, John J. <u>Vietnam Studies: Airmobility, 1961-1971</u>. Department of the Army. Washington D.C., 1989.

Wensdofer, John. "Letters Home, May-June, 1970." Charlie Company, 5th Battalion, 12th Infantry, 199th LIB.

After Action Reports, 5th Battalion, 12th Infantry, 199th Light Infantry Brigade, S-2 & S-3, May 1, 1970 to June 30, 1970.

After Action Reports, 2nd Brigade, 1st Cavalry Division, (Airmobile), May 1, 1970 to June 30, 1970.

INTERNET WEBSITES

Ovitt, Bert. http://www.minwest.com/warriors/ (5th Battalion, 12th Infantry, 199th LIB).

Ward, Tom. www.redcatcher.org Official Website for the 199th LIB's Redcatcher Association. If you served in the Brigade at any time from 1966-1970, please visit the home page and sign in).

http://www.army.mil/cmh-pg/books/Vietnam/Airmobility/airmobility-ch11.html

http://25thaviation.org/id261.htm
25th Aviation Battalion, Vietnam

http://www.army.mil/CMH-PG/books/Vietnam/mounted/chapter7.htm

http://www.vietnam.ttu.edu/
The Virtual Vietnam Archive
II Field Force Commander's Evaluation Report
Cambodian Operation 31 July 1970

www.vietvet.org/glossary.htm
Website for Vietnam Veterans

ABOUT THE AUTHOR

Robert J. Gouge lives with his wife, son and daughter in the mountains of western North Carolina. A serious student of the Vietnam War and the 199th Light Infantry Brigade, Gouge is an active member, deacon and teacher at Newbridge Baptist Church and a U.S. History teacher at A. C. Reynolds Middle School. In addition to writing and researching, Gouge is also an avid outdoorsman and enjoys traditional black powder hunting and shooting.

73156307R00126

Made in the USA
Columbia, SC
02 September 2019